1-800-FOR-VETERANS-ONLY

inspirational writs by

CHRIS BENT
www.chrisbent.com

Published in the USA by Chris Bent
Naples, Florida USA
http://ChrisBent.com

1-800-I-AM-UNHAPPY,
1-800-FOR-WOMEN-ONLY,
1-800-LAUGHING-OUT-LOUD,
1-800-OH-MY-GOODNESS,
1-800-FOR-SEALS-ONLY,
1-800-OH-MY-DONALD, and
1-800-FOR-VETERANS-ONLY
are trademarks owned by
Chris Bent and are used with his permission.

———•◦•◦•———

Also By Chris Bent

Available in Paperback and Electronic Versions

1-800-I-AM-UNHAPPY
Volume 1

1-800-I-AM-UNHAPPY
Volume 2

1-800-FOR-WOMEN-ONLY

1-800-LAUGHING-OUT-LOUD

1-800-OH-MY-GOODNESS

1-800-FOR-SEALS-ONLY

1-800-OH-MY-DONALD

DEDICATION

To Christina, Candice, Courtney and their journeys. . .

. . . and all Veterans out there and their families

. . . and the Teams' Brotherhood.

Prologue

This is meant to be a book for just one person. If just that one person is touched in some way to make their journey better, then the effort is not in vain. Each one of us can look back to one moment that changed our direction for the better. May this book, a collection of my writs and wit, find that pair of eyes.

Chris Bent

Kennebunkport
October 2015
www.ChrisBent.com

Contents

Getting There

Being There

Leaving There

Last Battle

This is the 8th book in the 1-800 series to date. It includes some chapters from previous books and many new ones. I was humbled during the whole journey. I don't know why I wrote this . . . It seemed like I was being carried along by forces beyond my control. . . .

I am a veteran. I have not had the pain and disappointment so many others have endured. This is intended to serve the veteran who reads this. Perhaps a chapter will be a stepping stone for some.

I have had the humble honor to have been a member of the Naval Special Warfare community. I was an officer in UDT-21, Underwater Demolition Team 21. UDT-21 became Seal Team 4 two decades later. From Hell Week to Apollo and Gemini recoveries it was quite a journey. Never in combat, but life has a way of challenging us all in other ways. I am humbled before my brothers who have been the tip of the spear and every veteran who dared to be one.

Veterans tend to be silent about their journey.

Veterans deserve more than they are getting.

I hope this book helps.

Thank you for your service.

God Bless America.

1-800-FOR-VETERANS-ONLY

inspirational writs by

CHRIS BENT

A Veteran's Comment on the Chapter "The Hand"

"I agree because when this USMC veteran returned home there were no handshakes or high fives but plenty of shaken fists.

I'm reminded of a verse from "Where No One Stands Alone"

"Hold my hand all the way every hour every day
 From here to the great unknown
 Take my hand let me stand
 Where no one stands alone."

There are two photos of hands representing two distinct eras:

The first is a stained glass window in a Chapel at Paris Island. S.C., with The Hand of God holding 12 Marines from my unit who were killed on Jan. 20, 1968 in Quang Tri Province, Vietnam.

The second is a marble work entitled "Hand in Hand" that stands at the entrance of a children's rehabilitation clinic in Dong Ha, Vietnam, just a few miles from the site where the above Marines were killed.

Floyd Killough, USMC (Ret.)

Getting There

The Patch

It was up on the wall.

Like a fly.

An embroidered WWII army patch.

Maybe the 2nd or 8th Infantry Division patch.

Maybe I was 10 years old.

What is it about the military that captures a boy's or young man's attention?

What is it that deep down beneath beckons?

A story from an old veteran that makes it all the more mysterious…. And compelling.

Maybe it is the video game of today. Where incredibly realistic good guys kill the bad guys. Blood is left on the screen as one goes furiously forward to score more blood points. Victory by the good guys. Never quit until you win and learn the game.

Different things attract one in life. From beauty to misfortune, from celebrity to saint, from good to evil. So many impulses to address. So many temptations to avoid.

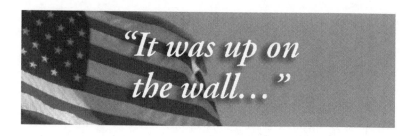

"It was up on
the wall…"

The real soldier knows. The veteran knows. Really.

Maybe you see an air show, or maybe you go on board the USS Intrepid in NYC. A young boy or man will have eyes opened wider, consuming more than he can assimilate.

The patch is on his wall.

The curiosity is imbedded.

Every veteran knows.

God Bless America

War Movie

Aren't they the best?

Nothing like eating popcorn in a dark theatre watching a war flick.

12 O'clock High. Bridges of Toko Ri. All Quiet on the Western Front. Deer Hunter. Apocalypse Now. Lone Survivor. American Sniper, etc. And for me in 1951, "The Frogmen".

War Movies really trigger the hidden man. His yearning to be somebody, to be a man, maybe a hero. And to be strong and proud. To prove yourself in a battle beyond a video game. Every war movie is now available on demand.

But there is a point where the fantasy will no longer suffice. Where the military offers the only option to realize the dream. Fighter pilot, Marine, Soldier, Sailor; each has an appeal. And all sorts of specialties abound. The experience potential is mind boggling.

There is one catch to the war movie, and that is, if you are in it for real, you may see and feel blood. You may see and feel injustice. You may see and feel brutality. You may see and feel tears. You may see and feel camaraderie.

Movie stars have all their lines written for them. They all are told how to

"*War Movies really trigger the hidden man.*"

move in a scene. Their uniforms and hair are planned. Lights, camera, action. Except it is a joke.

You come out of a movie feeling excited and manly...or sad...but alive. And that is the point. You never were in danger. Never. Only in the game the screen played with your mind.

War movies will get you if you watch them.

Most people go through life never enlisting for Real. Discomfort avoidance defines their manhood. Image and poise based on self-deception.

The real veteran does not have to open his mouth. Whether in combat or not he is the silent hero. His life was there to be taken by circumstance in defense of his country and your family.

You can go all over the real world and come back knowing what you didn't.

You can make friends you never thought you could.

Your country needs you.

A veteran makes a difference.

Top Secret

You will never know.

A veteran's lips are sealed.

And the only way you can find out is to be a veteran.

A vet never discusses what is Top Secret.

Now the media reporters and writers are constantly trying to peek under the rock. But it is too heavy for them to be able to discern the truth. Headlines abound with snippets of secrets, but you can't tell the false from the real. Smokescreens hide the truth.

So forget about finding out about Top Secret unless you are willing to enlist.

There aren't many places where a vet can share these secrets. Only with special people at special times and in private.

Top Secrets can be about weapons or covert activities or intelligence. But the greatest secret is what the Veteran really thinks. They don't really share what is deep within very often. You can't even begin to tell a spouse as they do not have the vocabulary and specific experience to understand. It would not be fair to them. So the vet will go out and

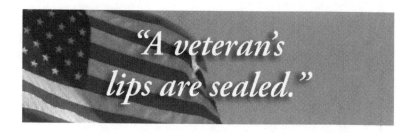

*"A veteran's
lips are sealed."*

laugh and have his beer but keep his real secrets safely kept within.

From enlistment to discharge the stories are fascinating and humbly powerful. You can ask a vet, but you will never know the right question. You may skim the surface but never hear the richness of the detail of a million special moments.

You can never tell what a veteran has seen by his looks, unless the injury is visual.

He won't talk about it.

He has been there for his buddies, platoon, and team.

A top secret privilege.

To fight for what is right.

Don't Go

Don't go!

The mother's heart screams.

Classmates in school separate you out.

You are no longer one of them.

They return to the battlefields of their cell phones.

The call of the wild is the call of fairness. You know there are guys hearing the call. They inhabit the pasts of all nations. The fight for what is right must be fought by someone. It is noble and it is just.

Once you go you are a hero, just for going. Where most avoid. As a little kid you looked up to anyone who had been in the military. Whether they answered phones or flew a fighter they were special.

The veteran does not look at himself as a hero as he sees everyone around him as better. But he is wrong.

Don't go.

Hey, when someone says not to do something it begs to be done. "Don't go in the ocean"…. And you do. And you see what others who don't go don't see.

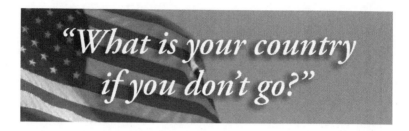

"What is your country if you don't go?"

You have to go. The real world is out there. Impoverished people. Brutalized by a total lack of fairness. Hungry. Lost.

Walk away and what does it say? About you?

All your life you hear don't do this and don't do that. It keeps so many from doing anything. What does Nike say??? "Just do it!"

Life is not seen on a video screen.

How can you know what someone is talking about if you don't speak their language?

How do you hold the hand of the lost unless you are there?

How do you feel the tears of the mother holding her dead child unless you are there?

How do you become a man if you don't risk?

How do you become a man if you don't go?

What is your country if you don't go?

The Flag

Red, White, and Blue.

Stars and stripes.

Tombstones white.

Honoring fields of green.

Was it worth it? If these dead did not go what would have happened? Some say a horrible waste. Some say they saved millions from terrible fates. Evil does not negotiate. Evil is very smart and loves weakness and lack of resolve.

Stupidity and self-absorption does not recognize evil. These are the true murderers.

Our Flag flies high. Stars added as the union grew. All over the world young eyes, seeing it for the first time, see hope and freedom. It is taught to some that it is evil. Shame on those teachers. They are the defilers of logic and good.

The Flag is saluted and raised with respect across the nation. It is the symbol of what the white tombstones honor. No Flag, no freedom.

"Beckoning the brave to serve her."

New freedoms today. Cell phones now link all minds for instant thought check. Hmmm?... If evil were not checked our cell phones would be taken away.

Our country is a nation of values and is not to be defined by her communication tools alone. The new generation will surely fight for them. Freedom of speech is now a deity. Don't tread on it.

Rolling waves of gravesites. Peaceful serenity. Tears of love. Unseen.

The Flag flies unfurled in the breeze.
Beckoning the brave to serve her.
And to serve all those who look away.

Red, White, and Blue.
Stars and stripes.
Tombstones white.
Honoring fields of green.

God Bless America.

The Uniform

PFC. Private First Class.

SN. Seaman

As soon as you put on a uniform things change. Forever. Regardless of rank you are looked up to by the kids on the block. Forever.

Why does looking uniform in a uniform seem so special? Everyone looks alike. Everyone moves alike. Everyone talks alike. Everyone appears to be without freedom. There can be confusion, then the command of "Attention!" brings everyone to stand in defined rows.

Salutes seem to be so special. Paying respect to what? The visible act of paying respect resonates deep within. As a kid you were always noticing acts of respect. Did you respect your Dad and Mom? You sure did if they were okay people. They gave you unconditional love. Gotta respect that. We want to respect people, we want to respect good, we want to respect good rules.

Expect respect? You know when but you never ask for it. You respect the uniform because it unites. You respect rank because it recognizes experience.

Can you imagine a ship with hundreds of sailors all wearing what

"Wear it proudly."

they want and changing every day? Visual chaos. The uniform allows everyone to be equal. Sure some get promoted that shouldn't, but that is life. So much can be learned in the military, in the wearing of the uniform.

On the sleeves go rank and length of service. On the chest over your heart go the ribbons of journey and merit. These are over the same heart that gave you the courage to join and the courage to care.

An Admiral or a General gets to wear stars. The same stars on the Flag. On camouflage uniforms rank is hidden so it will not be an easy target.

Weapons, armor, all kinds of equipment hang from uniforms when sent in harm's way. Uniform patriotism allows the soldier to step out into the night. To face the unknown and manhood.

The uniform gets you there.

Wear it proudly.

So kids can see.

That you'll fight for our country to be free.

Enlist

Is he nuts?

Whisper the peers.

Thinks the mother.

Not the father.

The recruiting office in the strip mall beckons every time a young man's glance tries to see inside. Army, Air Force, Marines, Navy. Posters crafted with drama. A wall of heroes. Escape from the video game. Escape from the small town. Escape from the city. Escape from unfairness. Escape from inner boredom.

The day comes…. After the birthday, after the graduation… the day comes and your put your hand on the door and pull it open. Bigger than life is the recruiter all ablaze in his uniform. "How are you doing son?" and your hand quivers and rushes to ink your signature. Save that ball point pen, it is the first medal of your journey. A symbol of your courage. Steal it if you have to.

You actually get to pick what branch of the service you want. And you know nothing. If you join the Navy you can get on a ship and see the world. If you join the Army or Marines you will learn how wonderful

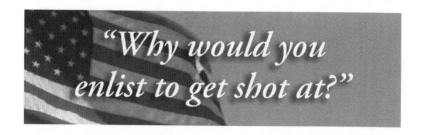

"Why would you enlist to get shot at?"

walking is. And the Air Force? Well… few really get to fly.....

Why would you enlist to get shot at? Why would you enlist to see friends get killed? Why would you enlist to get yelled at? Why would you enlist to get uncomfortable? Or miserable?.....

You are starting a journey that separates you from the pack.

The pack knows it too.

Your family knows it too.

Pride is hidden in their love for you.

Your country is welcoming your decision and you.

All of a sudden the Flag waves to your eyes with new meaning.

Red, white, and blue is the new true of you.

Are you nuts?

God Bless America.

Boot Camp

Is he nuts?

So you get on a bus, or train, or plane.

Looking out the window lost in thought and apprehension as you leave your life behind.

Bus to the boot camp of your choice where choice is no longer a part of your life.

"Fall in!" screams someone you never have seen before. You are berated as being essentially stupid and of little worth. All notions you may have had of self-importance are irrelevant. But it does feel good. Someone is going to take you somewhere no one else could. "You've got a ticket to ride…"

I suggest you are lean and mean so there is less pain to bring on yourself. Get to know the ground because you are going to live in its dirt, and then be told to have your uniforms perfect in a half hour.

Learn to love to run. It makes walking easier. Pushups, pull-ups, sit-ups will be your new breakfast cereal. The breakfast of champions. In fact you are being prepared to survive and to win. As your character is remolded around rules, order, values, and respect.

"So you get on a bus, or train, or plane."

Boot Camp. Where did the name come from?? Because your boots do all the talking. You learn to appreciate your boots for they can save you. You learn to polish them to see your face. Dirt can always be transformed.

Recruit trainers knows where you must go to be of value. They own you. A new family. A brotherhood second to none.

After enduring all kinds of seemingly nightmare conditions in rain, mud, wind, heat, and cold you find a new pride.

Graduating from boot camp is just a beginning. But what a beginning. And the end of being a kid.

You get a uniform.

And at home you are a hero.

But you know otherwise.

You just didn't quit.

First Duty

The bulletin board posts the list of recruits' first assignments.

Where you are going to be sent. Wow.

Some are going on to schools.

Some are going to units on bases.

Some are going to ships.

All who you were formed with will never see each other again.

Your first duty will be new faces and new authorities and new rules and new histories. You will work with men who have seen evil. Attitude will be everything. Enthusiasm will be rewarded.

New technologies abound. Amazing groups of guys doing amazing things. Weapons are put in your hands as if it was normal to have them. The game becomes real. Your ears open and you begin to hear about real.

Immediately you belong to a new trust group. From all walks of life. You get to see through their eyes as your sight and understanding of life grows. Your Petty Officers, Sergeants, and Officers are all there to teach and protect. Wow.

"You will work with men who have seen evil."

Many will leave port, many will fly for places not known and be immediately immersed in a different world and different cultures. Watching a sun set far out at sea or seeing it rise above a mountain or desert in a place where you cannot pronounce the name….. is humbling.

You want to make a good impression. And it pays to do so. You are now a veteran. All veterans know exactly what you are going through and what is ahead. Whether you stay in for a few years, or many, you are a veteran. You already have a story. In fact, most days up to this point have their own stories built in if you keep them. Many laughs through tears of unbelief that you were asked to do 10 or 100 pushups for doing nothing. That is a story for those left behind in your home town. You may only be a veteran of 100 pushups. But you are a veteran.

First duty is your first time to show yourself after high school or college.

It is your duty to protect your comrades.

It is your duty to be someone the young can look up to.

It is your duty to your parents to do good.

Because you will be fighting for good.

World News Tonight

The ABC anchor on the 7 PM World News Tonight said there were hostages taken in the country of "whatever".

In his perfect hair and suit and tie he looks us straight in the face and describes the plight of the world.

He looks comfortable and safe.

He has a commanding voice.

Meanwhile "whatever" is happening "wherever". And "whatever" is happening to guys uncomfortable from never knowing if their next step will be their last. The food is so-so. No restaurant to go to after the broadcast. Chow hall maybe…. Or the Mess if you are on a ship. Aboard you get to pass one another sideways as the corridors are narrow. Bunks or cots or hard ground is where most sleep when they are "wherever". No AC or heat in the deserts and mountains.

30 minutes and he is done. You are not.

The anchor waits until the next day to hear who has made it through the night, and what IED surprised who. The grief of families handled in 15 seconds.

"Every veteran has a fascinating story."

Jets race through night skies and ordinance expended makes for great clips.

Ships pierce the waves and storms to get closer to "wherever" and fulfill their tasks. Some sending men silently into the water in the night.

There are so many stories that are not told. Every veteran has a fascinating story. It is truly amazing to see an old guy describing something so vividly that happened a lifetime ago. Honored to be alive and honored to have served.

Stories that can't be told on '60 Minutes' and they aren't controversial enough anyway.

It is time for a commercial break.

These days many.

These days longer.

And the commercials pay for the news.

So do those who shed the blood.

Being There

Tall Tales

It's time for stories.

Where do you go to get them?

Too much beer and they are tall tales.

And the buzz demands embellishment.

Where lies the truth but deep within? Most veterans are hesitant to talk as they think little of what they did. They all do. And that is as it should be.

But these stories are the rough poems of life. Dirty, grimy, sweaty, exhausting tales of moments. We almost can tell them with a look. Most don't write about them as they feel they are too unimportant. These are the ones that are uncut gems. The motherlode of life is in the mind of the veteran.

If you can ever get a veteran aside in some private place and time, ask him. Listen carefully so you can create the next question. We love to share, but not brag. So we keep quiet. Those who knew combat don't want to relive anything unless you are very special. But then what you will learn will be special.

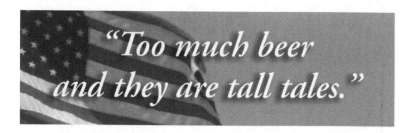

"Too much beer and they are tall tales."

Some of my BUD/S 31E classmates saw combat. I did not. Some went to Vietnam. One ran Seal Team 6. We are long gone from one another so I have had no time to share. Then there is such clutter in every moment that it is helpless.

Spouses and children are everywhere.

So silence is king.

I am going to share some stories already written that are in my other books. But these are not the same as the profound exhilarating sharing I have found in my church ministry and in assisted living homes with veterans.

Later on we will explore the hidden potential of the veteran, his perils, and his post-service journey.

The Drill Instructor

The Drill Instructor (DI) barks out his commands to the frightened cadre.

"Gimme 50!!!"

"Backs straight!!!"

"Chest must touch ground!!!"

"Full extension!!!"

"Get that back straight!!!"

"How many times do I have to tell you?!!!"

"OK, stay at lean and rest until I come back, children!"

Do you think they make Marines, soldiers, much less SEALs by asking them what they would prefer to do? Do you think the Drill Instructors have graduated from sensitivity training? Are Democratic Constitutional rights in effect? Are the recruits entitled?

I postulate that this just might be about life and death, not feelings. I think this may be about protecting your freedoms and right to complain. But this cannot happen without young men prepared for battle and for death.

"How many times do I have to tell you?!!!"

Huuuhhh??? Hello?? You have got to be kidding? OMG you can't be serious? Death doesn't happen other than on the news and usually far away. Whatever. Text me later...I am busy.

I am of the opinion that every young able-bodied male should be drafted. Boys become men much sooner than on the playing fields or the streets.

We sure don't like systems which tell us what to do. It is like it is in our DNA to rebel against authority. Go away Dad, go away teacher, go away policeman, go away boss, go away reality. I am armored with my self-serving iPhone and the courage of my social network.

We allow the Drill Instructors of life to make us miserable as long as it keeps us from getting killed. We love our bodies as they are our ego cathedrals.

But what about the Drill Instructor in the clouds? He wants to save our souls and enrich our lives beyond our wildest imagination. Problem is... we don't have any imagination.

How many individuals choose paths right into the combat zones of greed, lust, envy, pride, gluttony, anger, and filth? Drugs, alcohol, sex, vanity, indifference, and jealousy and... on and on. "Well, I didn't think it would be this bad. A little more won't hurt..." Consciences trampled by disdain for authority and distain for logic.

Where is the Drill Instructor when I need him? Everyone doesn't screw up their lives. I see some happy people around… No, not the hypocrites… just a lot of normal people going about their business, helping others, and laughing.

How do I become one of them?

Well, duuuhhh???

Hello, The Drill Instructor in the sky!

No pain no gain they always said…

Instructor Waddell

I wrote a piece a while ago entitled PDTMWTD…

Please Don't Tell Me What To Do.

It was targeted at kids who don't like their parents telling them what to do… Well… and maybe it applies to husbands too???

You get the picture… any of us at any time.

PDTMWTD.

When we tell someone what to do we are trying to warn them of possible consequences that we see that we think they don't see.

We all have to learn the same hard way; by experiencing and dealing with our own decisions. We want them to be ours. Why should we trust anyone else's opinion? "Enough with the advice!" we say. Of course, I am still saying that to my wife today….

Then… when we get a job we have to listen to our boss. Groan. If we really grow up we listen to our customers and clients, not ourselves. This is difficult, but it happens because it is paycheck driven. We listen and learn.

So it appears it is the incentive that is important. What is the incentive today when all kids withdraw into the self-assuring world of their cell

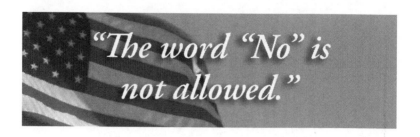
"The word "No" is not allowed."

phone? We are in trouble. They are affirmed by their insecure social network, not by our wisdom. They no longer "look up" to anything.

The word "No" is not allowed. Children are taught the nuances of "child abuse" and only have to tell their teachers and there will be a knock on the door.

I was blessed with great parents, spankings, and a plethora of "No's". Thank God.

But it was not until the military that I began to fully appreciate what "No" meant. If you wanted to be a Navy Frogman/SEAL the INSTRUCTORS were waiting with snarling teeth and monstrous threats of pain. If you did not say "YES INSTRUCTOR" to everything they suggested there was a real price. Too many pushups, too many sit-ups, too many pull-ups, too many runs, too many swims… "Are they crazy?" I would think to myself. Is this nonsense really worth it? Why were they so strict? Why was my instructor Bernie Waddell so scary behind his wry smile?

Why??? Because all hell could rain down on you if you showed disrespect in any way, shape, or form. So just do what he asks and get on with your moment. Mile after mile crawling, running, and swimming in the freezing winter during Hell Week weeded out the non-survivors. The Yes Men.

You see... the instructors knew what would happen to you if you were not forged properly. You would be killed when it could have been avoided. Your swim buddy would have been put in peril unnecessarily. The mission aborted because you did not heed your instructor.

We need instructors in order to become all that we have the potential to become. Values must be honed in sweat and pain. We need to learn that freedom is not for free. If you are not conditioned, you will not be strong enough to stand in the winds of life. Laziness destroys potential.

There is white and black, good and evil, yes and no, do's and don'ts....

Values became valuable.

Serving others became more important than serving self.

Prayer became real.

Morality was affirmed.

God Bless You Instructor Bernie Waddell... and Godspeed.

And God Bless all the instructors everywhere.

And parents too.

Sea Legs

Sea legs is an old nautical term that describes someone new coming on board and the time it took to adapt to the rhythm and rolling of the ship in heavy seas.

It is, at first, tough to maintain your balance when every axis moves.

There are smells and sounds all of which can make one seasick.

Heck, even some of you landlubbers can get sick on a cruise ship doing nothing but sailing smoothly along. I remember some very stormy rough plunging days on a small ship in the pacific. I was new, and up on the conn it was a roller coaster. But the throwing up soon ceased as I got my sea legs.

We all think we know more than we do. You think you know as much as your boss until you have 10 years under your belt and you realize there was more to learn.

Try and tell teenagers they don't know what they are talking about. They have sea legs for nothing.

Now one would think cell phones are so easy. That facebook is so easy. That twitter is so easy. And on and on. But even these require sea legs!! Over time one makes mistakes in saying something one shouldn't have

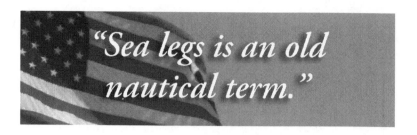
"Sea legs is an old nautical term."

to someone you shouldn't have. Copying someone by mistake. Several times I wanted to throw up. I wasn't careful about what I was doing. I didn't have my social networking sea legs!

It is human nature to assume one knows more than one does. As you get older, and as you get really older, you see things you never did 5 years prior. Aging is kind of like getting life sea legs.

How about what one knows about what is right and wrong?

How about what to do about evil or…. Unfairness?

How about knowing when to just act without further discussion.

How about setting one's own lines in the sand?

In the sea of life, sea legs means knowing that helping others is the only thing that can keep you from heaving over the side……

EPCER 857

USS MARYSVILLE EPCER 857.

They sent me there as my first duty station in the Navy after college and OCS.

It wasn't what this 22 year old wanted.

Sometimes we have to do things we don't want…

You couldn't refuse or say "whatever" back then.

Experimental Patrol Craft Escort Rescue….. She was stationed out of the US Naval Electronics Laboratory on Pt. Loma, San Diego. I came aboard on Jan 1, 1963.

You see, the Navy had decided that potential Naval Special Warfare Officers, UDT/SEAL, needed some real shipboard experience before they became "special". Reputations as cowboys and independent behavior needed a solution.

So we towed a 500 foot thermistor chain at 5 knots between Acapulco and San Francisco for a year getting temperature versus depth data. Slow and boring?? Not really as you were always busy on the 184 ft. steel ship. Some big storms way out there all alone. Learned a little about how Navy things worked. It was good.

I left and reported into UDT/SEAL Training, aka BUDS on Jan 2, 1964. Winter Class. It got cold. The Chesapeake got nasty. Hell Week was nasty. It all was such a shock that you didn't have time to digest food or what was happening. I wouldn't quit. I learned to laugh and swear and to trust.

But this is about having to go places you don't want to. This is about learning from every experience. Every experience makes you more unique. You won't know until years later what it was all about. How you were forged by good times and bad times. When we are young we are so ignorant and arrogant. Our opinions are so off base (pun intended) that it is a miracle we survive.

Maybe EPCER 857 taught me things I needed to know. Maybe I would have quit BUDS without that experience. I have no clue. But do any of us?

The lesson is to forge ahead, one day at a time, sorting right from wrong.

Easy from hard.

Not quitting.

Finding trust.

Taking orders.

Following rules.

Holding out your hand to the one reaching towards yours.

Always willing to help.

5 Knots is not so bad if you get where you are going and bring back something of value.

The Camaraderie

It is the camaraderie.

One searches for truth and meaning under the guise of what makes one feel good.

There are entertainments, friendships, and all kinds of things which appear to be what life is.

Often these are devoid of purpose, just activities that make one feel good in the moment.

But, "Why am I here?" echoes deep within. Where do I make a difference? This is the man-quest. Women have birth and family. Man searches. Man succumbs to aloneness and journeys that lead nowhere. There is pain without purpose.

Military veterans come home and bring their private pain and surface smiles. Many have seen the belly of the beast. But it is the basic training and living together as brothers with tomorrow's uncertainties that forges a bond unimagined by civilians, by families even… Eating and sleeping often in extremely difficult circumstances. Never knowing the closeness of the unseen enemy. There is always a mission and purpose to the next day… even if it is just to take care and be with your buddies, your brothers in cammo.

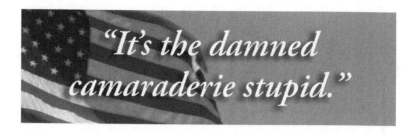

"It's the damned camaraderie stupid."

The meaning of your life is defined and private prides are worn.

Then you embark from the returning aircraft and receive the hugs of family. No one knows what your eyes have seen. They want you back as before. You are not the same. They can't get it....only your buddies who are spread far and wide from states to cemeteries.

Finding a job is often hard. Employers have no shared experience. They have no clue as to the disciplines and focus you have experienced. They are the amateurs in life, but don't know it. You have graduated, but don't know it.

Manhood comes from the camaraderie of combat; of getting close to evil. From being next to a buddy's last breath. From touching an artificial limb.

One had learned to communicate without sound. Eye contact rules...

Caring for a buddy more than one's self.

It's the damned camaraderie stupid.

Hire a veteran.

You owe it to yourself.

Only Easy

At BUDS there is an expression; "The only easy day was yesterday."

It's really about life, marriage, work, family, and fighting for what has value.

What that is, is for us to decide.

But if there is nothing you will fight for then you are nothing.

When you are young and strong little seems impossible. If you are in Hell Week and cold, nothing is probable. The moment must be seized with resolve, humor, cursing, and caring. You won't quit if your concern is the men around you… your new brothers in pain. One exhausting hour after hour becomes an investment in the future. An investment in making it to Friday. Each hour is an accomplishment to be proud of. You build on it and feed from it.

Don't quit on your family and they will know not to quit on you. Once yesterday is yesterday the pain is just a passing event. It no longer hurts. It was easy.

Today, in the moment, is the challenge. Attitude shapes all. Blood is a badge. A smile from your misery mate is a badge. Smiles affirm pride, the right kind. The heads lower and you get on with the next evolution.

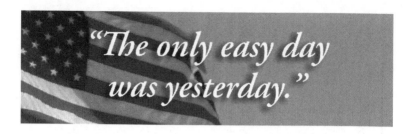

"The only easy day was yesterday."

No cell phones to get assurance from. Too bad and too late. Your only phone is your smile.

Isn't a smile so fantastic when you catch one from a person under great stress? Coping by seeing some humor in a desperate moment.

"The only easy day was yesterday." bonds thousands of guys who know what it means. It is worth it to find out what it means first hand.

You just have to know how to run and swim easily. Do not wait until the last minute or the last year. Learn about the ocean…. Surf…. Learn how to fire a gun… go out in the woods and get lost….. and figure it out.

Don't be easy on yourself.

Life isn't easy.

The unexpected is to be expected.

Be proud of yesterday.

Make today a great yesterday too.

And… keep the Faith.

Running Lights

Would you believe that today joggers go out running at night with small LED flashing lights?.... Like giant fireflies?

Men don't run at night with lights. Who ever heard of such a thing? It should be women only... Then you know who to help... LOL, wink, wink...

Now cars are mandated to have daytime running lights, DRL. Bright halogen LEDs in every shape that may tell you what kind of car it is.... Crazy times...

Now all you sailors who love the sea cannot go out on your own if you don't know the Rules Of The Road... Or all the nautical traditions that define behavior and right of way at sea.

From sailboats, to motorboats, to yachts, to cruise ships, to freighters, to aircraft carriers... all behave by the same Rules of the Road. At sea everybody has a strict code because if there is an accident you not only get hurt like in a car but you sink too... Much more pride is taken by sailors than drivers....

50 Years ago on a winter night in the Chesapeake, Dan and I had just slipped over the side of our small LCPL boat breathing pure oxygen

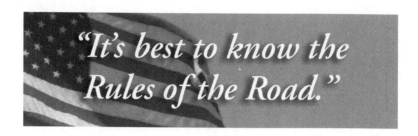

"It's best to know the Rules of the Road."

from our Emerson Rebreathers. The target ship was moored about 2 miles away. We took our bearings and started the swim.

After a while I heard this thumping in the black. Thump, thump, thump. We should not have been in the shipping channel. I thought it may be the propeller cavitation of a large ship so I surfaced slowly to take a peek from the top corner of my mask so as not to be seen regardless.

I saw the red and green running lights of a large freighter high off the surface. Running lights are configured so if you only see red then you are looking at her port side. If you only see green, you are on her starboard side. BUT, if you see both she is heading right at you! What to do and to do fast?

I said to Danny, let's go to the bottom at about 50-60 feet and let her pass over. We started down... and would be diving deeper than 33 feet where oxygen becomes toxic. And... also envisioned being involved with the propeller if we erred. I reversed the call and we took off for the surface, figuring we could use the bow wave to separate us from the hull and screws if we were that close. Better to see what is going to happen to you.

She got closer and closer looming above us and... we were spared as she passed within 20 yards to port.

We continued our swim. And smiling, hit our target and rose to the surface with another "Gotcha!" for the lookouts who never saw us.

It's best to know the Rules of the Road whether at sea or in church…..

Chicken Neck

Do you remember back when you were young and others dared you to do something?

Or when you dared someone?

Immediately, if there was hesitation… the label "chicken" was thrown on the emotional table.

Nope, no one ever wants to be called a "chicken". Especially SEALs…. LOL

Well, Marines too, and all the rest.

I guess I can't call a base jumper "chicken". Jumping off mountains to free fall…. Or even Astronauts. Or even Pastors taking on the cynics…. Nope… no chickens there… Or even mothers raising children, much less doing it alone….

I remember an afternoon after Hell Week when we were taken to the woods to camp out before doing an overnight demolition raid on a bridge 10 miles away. We had to arrive at dawn unseen to position our dummy charges. No clue what the night would bring, stumbling with precision through the black wilderness.

We were each given a shelter half (Aka half a tent). So my tent buddy and I put it together and prepared to get some sleep. But dinner first. We were hungry. (You are hungry all the time during Hell Week training) We had a pot of water, some potatoes and… were given a live chicken. My sense of humor kicked in and I quickly gave her a name and started petting her, like a golden retriever.

Now… I had no farm experience. I did not know my tent mate's background. Well, it became the time to de-feather and boil the chicken, but …. it had to be killed first…. Seizing the moment I handed the chicken to Bob for the ceremony. To my surprise, he declined to kill it. I had no clue what I was doing, but with a big smile I grabbed it and twisted its neck off.

That was the last thing I ever killed. He went on to Vietnam and also became a Commanding Officer of SEAL Team 6.

I left the Navy after 4 years and in one month went from "high explosives to handbags".

Our paths diverged so incredibly.

I have no clue of the covert activities he managed in his career.

But I do remember that moment when a chicken neck was going to link us forever.

Life gives us all unique moments that are ours alone.

Cherish them.

They must be gifts from Somewhere?

The Weaver

There used to be weavers in every town.

Fabrics of every texture and color.

Rugs galore.

Now you can only see real weaving in counties like India. Still beautiful handmade art if you can find it. Otherwise it is all now computerized and a human never touches it. If you want real texture and quality there is nothing like hand woven to stir the senses.

Some things, to be real, have to take real effort… and real effort by the individual. What is simple is always complex if it is meant to be the best.

There is a simple looking obstacle at BUDS. 10 foot 3 inch diameter steel bars parallel to the ground about 3-4 feet apart that create an inclined up and down design. You run to it then try to find an efficient coordination to weave your body into, under, and above, and under, and above…. Like human thread. Except that it requires real focus, rhythm, and effort to do it fast enough to pass. SEAL candidates trying to be a needle and thread at the same time. You are alone in your head and body. Once it snowed. You do the obstacle many, many times…

"10 foot 3 inch diameter steel bars parallel to the ground..."

especially during Hell Week.... And after. You get good at the obstacles. You have no choice. No time to text.

You learn that life requires you to weave in and out of complicated situations.

You take the word "No" out of your vocabulary.

"Yes Sir, Instructor Sir."

Good preparation for marriage???

"Yes Dear..."

"Of Course Dear."

Slide For Life

Water slides are fabulous these days.

They even have them on cruise ships… Go figure.

Kids used to pour water with a hose on the top of a slide in a playground and enjoy the wet ride.

Some big slides got dangerous… but still exciting. Who cares anyway when you are young.

A lot of people want to slide through life. So they throw a little alcohol or drugs on it and push off on the easy downhill ride. Except at the bottom the ladder up has rungs too far apart.

The easy way is always the hardest way. Until you learn that the hardest way is the easiest. We humans try to cut every corner there is. We cut too many moral corners. We try to maneuver around the Truth. Guess where it gets us? Not closer to happiness for sure…

Well, now we are forced to go back to the 'O Course' (Obstacle Course) to learn simple lessons. At the Amphibious Base in Little Creek Virginia and also in Coronado California are the only two Slides-For-Life in existence…

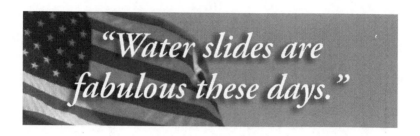

"Water slides are fabulous these days."

There is this 30-40 foot wood frame tower to climb up…At the top is a ladder to climb to then find a way to jump up at the rope that descends across a water pit for maybe 30 yards. But your body has to be on top of the rope to make any time descending. Let's say it is a leap of Faith to get to where you just start. Hey, it snowed too.

BUDS is all about faith that you won't get killed every time some outrageous task is put forth. Like running up a sand dune backwards.......

Life requires all of us to do the impossible at some point.

Faith is required.

Where do you find Faith?

That is what my five hundred 1-800 book chapters are all about.

They are damned good.

Look straight forward, never down, and the Slides-For-Life can be conquered.

Happy guys at the bottom will greet you.

Round The World

Thank God some bloke discovered the world was round.

Would you believe at one point they thought it was some cube and you could drop off the edge, drown, and cease to exist?

Haven't you ever felt that you were on some dangerous edge?

Financially?

Emotionally?

We travel all around the world looking at other cultures, their beauty, their history, and their plight. We usually look away from the latter. You learn a lot about life going round the world.

Sometimes our world is right in front of our nose.

For SEALs the world is their responsibility. They never know where they will be sent or if they can ever tell anyone… ever. Their world is black. The night is their domain.

I first heard the phrase "Round The World" on the third day into Hell Week. Sounded like fun. Except for the fact that we would be carrying IBLs, Inflatable Boat Large, on our heads for 24 hours as we paddled on the ocean, on lakes, and maneuvered through woods where we had to carry them on edge to get through the trees. Running, crawling... up sand dunes.... Lots of time for swearing and humor.... You had to be there.... LOL... obstacle course.... Stinky swamp mud..... and even the rock jetty. Did I say it was covered with ice you west coast surfers?? Talk about slippin' and slidin' folks... Shit, it was a masochist's blast.

All is a blur now, but I am sure many quit by Wednesday. We just kept plodding along... and along... try pushups in the sand with the boats on your backs.

The instructors conjured up all kinds of preposterous events that were just to scare us and prompt quitting.

But is not life the same...??

No matter what path you choose there are always curve balls. There are always opportunities to quit.

We have a saying... "Quitting is not an option".... Accepting that, what is one to do??

Keep on keeping on...

One step at a time.

Live in the moment giving it your very best.

It is easier when you are trying to help your struggling buddy, mate, or child.

But just don't quit and you will get round the world.

Now breathe deep, open your eyes, and say "I did it!"

Big Screw

I bet we all have been screwed at one time.

Something unfair was done to us.

Even robbed of something.

When you gamble you get screwed.

When you play with the Truth you will get screwed.

The big screw is coming.

It's in the Middle East.

My first big screw came in Key West.

We were at Underwater Swimmers School at the Navy Base.

That night, in pairs, we entered the harbor with closed circuit oxygen rebreathers. Our assignment was to attack a moored ship at the dock a mile away. It was dark and spooky. You stay close to your swim buddy. It's nice not to be alone, and much less dangerous.

Suddenly the water swirled violently as a school of Jack fish erupted around us. After swimming at 10 feet for 20 minutes a quick peek indicated we were way off course and out near the harbor entrance.

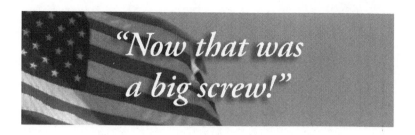

"Now that was a big screw!"

Took a compass bearing and slipped back below and swam on. Then it felt like we weren't moving. We had literally swum into the fine mud bottom where the silt density changed so slowly we didn't initially notice. Yuck luck.

Came up again for a peek and swam for another 20 minutes until I hit something solid. We were in the pilings under a pier a good 40 yards from our target.

There was one problem. A moored submarine was between us and our target. Perpendicular. What to do?

My logic said do not gamble going back out to the harbor but take a direct shot by going under the sub which would allow a straight course to the ship.

We go down, embracing the enormous circumference of the submarine. At her keel point she was only 3 feet off the hard sand bottom. Oops. If she decided to ballast and sink a little deeper we could be pressed into the bottom. We turned over and put our stomachs to the hull and swam. Then the straight course to our target.

Real dark.

My hands touched and groped around on something metal and big. It was a giant propeller, maybe 15 feet wide… We swam the propeller shaft to the hull and placed our dummy mine and rose to the surface with pride and jest to the surprised crew.

Mission accomplished.

Now that was a big screw!

How do we stop the one headed our way???

Fire in the Hole

Sometimes when camping as a kid we dug a shallow hole to build a fire in.

Maybe the forerunner of today's fire pits??

We see a lot of fire these days…

From forests, to brush, to CVS drugstores, to historical sites in the Middle-East…

Fire is something which consumes without remorse and without following anything but itself.

Sometimes emotions get out of control and rage like fires.

Is there fire in Washington?? Is entitlement a fire? Is playing with the Truth a fire?

Everybody in the SEALs knows the expression "Fire In the Hole!" It is something uttered just prior to igniting the blasting cap which ignites the detonating cord which then instantly detonates the 1,000 lbs. of C4 in an instant maelstrom of explosive force. Better keep your heads down.

My swim buddy Bob and I, and our men, had just laid an underwater

"Is there fire in Washington??"

field off Vieques of C3 of 500-1,000 lbs. and it did not go off. So the two of us, as officers, had the pleasant duty of paddling out over the live field and redoing the fuse assemblies. When done, I lost the flipped coin and dove down the 10-15 feet and pulled the pins. I think we may have had a 3-5 minute delay built in for us to paddle out of harm's way.

Fire In The Hole!!

As I surfaced, but unexpectedly into the bottom of the IBS (Inflatable Boat Small) rubber raft, I got disoriented and was slow getting aboard and paddling. It all went off… BOOM…. as water flew all around us. From shore we disappeared in a 200 foot cloud of water. Then smiles from afar as we paddled out of the mist.

Lesson learned.

If one is trained, thinking, and honest, the fires and explosions of life can end the same way.

You will be smarter and more careful.

Especially with the Truth.

PS: When I left the Navy in the summer of '66 I went to work for a department store in NYC on Fifth Avenue, Lord & Taylor. I get to say that in one month I went from High Explosives to Handbags! Yep, as an assistant buyer in the Handbag Department…. In the basement.

Fire In The Hole!

Get the Tan

I don't know for sure when it all started.

But sometime as a kid I met this UDT guy and he had this tan.

He also had this neck chain with something on it.

I heard they had some joint in St. Thomas called the Silver Bullet... where frogs and stewardesses congregated and did the Merengue ..... Tans were mandatory.

What is it about a tan? Girls sure look great in them.

Then there are the cruise ships of today. How do you walk the deck without tripping over a lounge chair? Why do hotels ruin the natural beauty with rows and rows of closely packed beach tanners? In Italy they have it down to a science. Good luck getting into the water.

I think the SEALs of today are so busy training and travelling that it is hard to find the few days to get The Tan. To them it is now a real luxury.

Who ever heard of a tan on a submarine? And the SEALs are the only ones to get topside ever so briefly before back in the water anyway....

A tan used to be a pseudo status symbol. Men who sailed and had yachts had them.

"What is it about a tan?"

But…. you can't judge a man by the color of his skin.

Our greatest sin is judging others, looking for easy ways to form an intelligent appraisal from afar. Color is often the first filter. It is usually downhill from there.

How can you judge the substance of a man without knowing his heart?

And without knowing His heart?

Life can bring you humility and wisdom only if you discard ego and pride.

Only if you learn that helping others is the first act that defines one.

And helping not just your peers but anyone, even the dirty.

Yes, you can and should judge and attack evil.

But learning how to judge takes wisdom, commitment, and integrity.

You will make mistakes.

You will be chastised.

Too bad.

We have to learn the Way.

WAG

Dogs wag their tails.

There is no guesswork figuring out why.

It usually means they are happy or anticipating/begging being fed.

Or best of all, you just walked into the room.

Nothing better than a dog.

I had this mixed Black Lab that went everywhere with me. Jumped off the boat if I did. Ran 4 miles in the dark at 5 AM with me. Went orienteering with SEAL candidates.... Could leave him outside on a run line all day... Slept near me. Warned better than any alarm. His name was Curtis. I miss him... I made this green neon sign for my office that says "Curtis Lives".

His wag was the start of all our waggers since. Puppy love.

But we are talking about the other WAG that is often as important. For sometimes critical decisions have to be made on the run... instantly. We call it, laughingly, the Wild-Ass-Guess. You have to be good at them or you will be dead.

It pays to train and train and train so you reduce the WAG factor to a

"Dogs wag their tails."

minimum. Yet, in the face of peril, someone has to offer the WAG for all to consider and act upon.

The sniper manages his breath, measures all the variables, but then the real menace of the human target often becomes a WAG…. And he "sends it".

It is important to assess all our WAGs so we are learning from them and reducing them.

I contend that a thorough involvement with the Bible is essential.

Find a version that works for you. If it is confusing don't quit.

Find a church or person who can interpret and make it enjoyable.

Don't get caught up in the memorizing folly.

It is the Spirit that is to be sought out.

It is harder and better than being a SEAL.

This is not a WAG.

Swim Buddy

Never get more than 6 feet apart.

BIG rule

BIG.

In the Navy SEALs, if you chose to ignore this rule, then on the next swim you would have a 4 inch 6 foot long hawser (rope) looped around your necks to keep you close together.

When you are swimming a mile or more in the ocean… and at night… and underwater… you need your swim buddy close so you can concentrate on your compass and navigating to your target. I preferred to keep my hand on my swim buddy's left hip, making sure he always knew where I was.

Swimming alone can be real dangerous. It was beat into our heads in our SEAL training. This is why I go nuts when I'm on a dive boat and everybody scatters once they are in the water. They have no clue. I do. SEALs do. Duuuhhh?

Marriage really demands the same respect. But we don't "get it" when we are young. And it is often too late when we are old. Hah! I bet you women want to really trust us and know where we are at all times. But, what if we want our independence and the thrill of a little flirt once in a while?

We should be going to church together. We should always take our kids to things together. Kids should know togetherness. They should feel you are their swim buddy too. Trust.

My wife drives me nuts wanting to know where I am every minute. Maybe that is the way it is supposed to be?

Maybe what they taught us in the Navy is really more profound. All of it was about trust. Going into foreign places where danger exists requires a lot of training, and trust...deep trust.

We all have our friends we think we trust. Some we have known a long time. Some talk behind our backs. You don't know who you can really trust until there is a real test.

One has to care more for the other person than self. You know it when it exists. The hawser teaches trust. It keeps you close.

Life is one big swim through all kinds of eddies and turbulent currents, even enormous waves that capsize you. When you have been sucked to the depths and come up coughing water and gasping for a breath, your swim buddy is at your side. Then you know trust. Maybe even love.

Kick, stroke, check your compass, check your depth, and that your swim buddy's hand is on your hip.

That closeness should there with your kids and…. Yourself.

That is real living, real family. All else is a delusion.

Ultimately, partners die or leave.

There is one swim buddy you can always count on.

You can read about Him in the New Testament.

American Sniper

American Sniper is a movie made from a book by a real person.

It is biographical non-fiction.

The author was murdered.

Chris Kyle was a good man.

A very good man.

His murderer was sick and evil and one of us.

This movie is proving to be one of the all-time largest grossing films.

All the competition are fantasy fiction movies.

All are pure entertainment and escapes from reality. Escapes.

American Sniper is not.

Its greatness and appeal is from confronting issues head on with artistic realism. We are forced into conversations that all other movies avoid. What to do about evil? Who can do something about it? Does evil exist? We get criticized if we try to define what is good or bad. Sad state of affairs.

The movie makes us think about what military families give up for

"Chris Kyle was a good man."

us. The movie provides moral dilemmas facing the soldier. It is not impersonal but personal. Memories haunted by acts of horror and doubt.

We have Navy SEALs to do some special assignments in the quiet. Risk taking at the extreme. Beyond conventional boundaries. Special Warfare is special. Training is beyond imagination. Others know nothing about it.

The movie misses giving us a real feeling of the exhaustion and pain endured. Hell week is a week of indescribable fortitude and compromise. You have to give up your being to trust. Trust that you will not die. You may quit at any time. A choice that terrifies you every moment.

Leaders are forged.

They know what it takes.

Moral leadership requires moral leadership.

The world needs us.

"Quitting is not an option."

Forever Silent

What do we know about what we don't know about?

How can we judge when we don't know the whole Truth?

But we do judge, and fervently claim we are right.

Do we have heroes we never know about?

The submariner's Silent Service?

The CIA operative?

The Green Beret alone out there?

The Stealth flight that never existed?

The SEAL returning unseen?

The private pain of a veteran?

The tear of a wife?

Heroes in humility.

Keeping us free.

"*The submariner's Silent Service?*"

Allowing us to smile.

Who pins the medals on them?

We will never know.

They want it that way.

God Bless America.

CO2

I have a friend, Philippe, who believes that CO2 is our number one priority.

Global warming and carbon dioxide are apparently the number one enemies of the moment. Hmm...

When I was a kid, you know... mid 20's... the Navy taught to me swim at night by using the Emerson Closed Circuit Rebreather. It gave off no bubbles so no one knew you were swimming under their ship to attach a limpet mine that would go off long after you had swum away. There was this chemical called baralyme, a mixture of calcium and barium that acted as a CO2 absorbent in a filter that was part of the breathing path. The problem was that if water leaked in then the baralyme would not absorb the CO2 which was toxic. Then you would rebreathe it and die.

So we did not like CO2. We liked living. Well, our planet does not like excessive CO2 and we should be finding ways to filter and reduce it. Let's have the North and South poles happy again.

At the same time I wish all this CO2 fervor could be channeled into all kinds of other issues like cruelty to humans. We have the SPCA. Where is the SPCH?

"It gave off no bubbles..."

Where do these politicians and leaders (choke) in all these other countries get the gall to watch their own poor suffer and die while they are having great meals every night???

Shame also on the many that treat women like animals.

Where is the SPCW?

Why are we and all our media not crying out in rage at these even more pressing tragedies? Why do we ignore the torture and abject denial of human rights and dignity in North Korea, much less the Middle East… and all the other places that are protected by political correctness?

Why?

Is CO_2 poisoning already creeping into our bodies??? Why is our judgment so impaired? What has happened to our values??? We don't have to bomb… but where is our outrage? Why are we always apologizing?

Come on America!

Do you believe in anything? Do we just withdraw from the truth until the next major war? How come our veterans and servicemen are so patriotic? What do they see that we don't? Why are they silent when they return…? Is it because they don't think we can understand? Or care???

They know values. Do we? Values are the cornerstone of our democratic society. They have been forged in blood and sacrifice for centuries. And now we act as if they were archaic principles not that relevant to our digital age. That our Judeo-Christian Constitution is old parchment riddled with holes and outdated irrelevancies.

Our problem is not with the environment, it is with us. We individuals do not make our individual stands. We are asking government to take responsibility for all our needs rather than forging them out of collective individualism.

Each of us squanders. Each of us is responsible. My dad always said, "Turn out the lights Chris." I do to this day. But we could watch all that we do. Use less energy. Become energy efficient. Become values efficient.

We need to manage our own details.

We don't look as healthy as we used to.

We have legislated so much out of what should be normal.

Where are the afternoon athletics in schools?

Where are the values that used to be taught in schools?

Let's take the blame ourselves in a courageous act of accountability.

The endangered is man.

We are the pollutant.

God help us.

Hurry Up

Who says it first in your family?

Hurry up dear!

Now I know we guys don't like to be told to hurry up.

I assume all women expect it as they have so much stuff to organize...

How about teens?? Are they ever ready? Are they ever on time??

Ever going on a trip and have your wife tell you to hurry up packing when she is telling you what to pack???

In the military everything is hurry up and wait. OMG at Ft. Benning we must have laid under trees for shade for a whole day in oppressive heat for a flight. Rush, rush, rush. Wait.

In life things seldom go as planned. But plans there must be as lives are at stake.

Actually having to wait can insure all ground is covered and last details re-sorted out.

Making kids wait is one of the best things you can do and teach. Wait for waiting's sake and wait for a purpose. Both are important. Teach them. On space launches there are built-in waits.

Wait until the Seeing Eye dog is fully trained? Wait until the broken leg heals?

Wait until the heart mends and feelings are back to normal. It is hard to wait, but it can be done with dignity and humor.

But the biggest wait in life is putting off what one calls "religion". "Religion" has become a bad word these days. Who wants to be associated with such a criticized position? So Christianity is almost being forced underground.... To wait???

No, to regroup, and find its correct path and then after the wait stand tall and sublimely obvious.

But this wait is not really about organized denominational religion. It is about what you feel deep inside about right and wrong. The journey to God is within. His Holy Spirit abides in each of us. It was there from birth. Look back at your first sense of fairness.....

We allow ourselves to allow the walls of hypocrisy, ignorance, insecurity, and denomination to rule our judgments. These walls give us superficial excuses to wait and put off serious reflection about God.

But when that time comes... and....

If you don't ever quit on Jesus…

Then you will find out who you were really meant to be.

Don't wait any longer.

HALO

HALO

High Altitude Low Opening.

It takes Oxygen first, then courage, then training, then a plane, then a cause.

Gotta have them all to be that stupid… at least to the earthbound.

Then you can add weapons, or explosives, or a rebreather and fins.

And then…. fingers that can still cross.

You jump out high… no…forget it…at high altitude…

Track miles to target area… Open low unseen. On land or sea.

No one knows you are there.

Your intel was True.

You get home safe.

When you leave your mother's womb she does her best to give you the courage and skills to navigate life. You start out with her love. Her values, her Truth. The highest preparation.

"High Altitude Low Opening."

However, most choose to descend to uncharted depths where vanity, pride, and insecurity lie.

Mom is gone.

The mother ship.

Dad had provided warnings but is often sadly gone when his stern discipline was called for. His Truth not learned.

So many open too low. They can be found regrouping in homeless and addiction centers... if still alive... Rehabilitating because they jumped before they were ready.

To navigate life requires certain wisdoms. The compass is Truth. It is not sold in a store. Some find it in church. It can always be found in helping others.

Values that have stood the test of time are available at the Jump School in the clouds. If you don't graduate you jump stupid.

And someone will say that was a dumb-ass thing for you to do.

Maybe old people have landed safely.

Sit down next to one and find out.

You won't be a "dumb-ass".

You get home safe.

Have A Lovely Opening.

"Stand in the door".

Long Swim

"Let's go ladies"!!

We got up real fast, real early.

Stuffed down a breakfast.

Grabbed our fins and masks and UDT swim vest.

Piled into a slow boat and departed US Naval Station Roosevelt Roads, motoring for several hours to the Island of Vieques. The sun was rising as we were told to jump overboard and swim back. Huuuhhh..???

Are they crazy? We have all said that haven't we?

The Caribbean water is warm... thank God. So it's just you and your swim buddy in UDT shorts and with a compass. Whew, that helps.

The water is clear and the coral bottom soon fades to dark as you kick, stroke, glide the SEAL sidestroke. Face to face with your buddy. Naked warriors. Sissy dive skins not yet invented. We were bare..... And nobody knew we existed.

There is nothing on the horizon ahead and soon nothing behind. You are in the ocean alone. The instructors are in a boat a ways away.

"How do you love and hate instructors at the same time?"

They have to keep track of all the swim pairs as they separate... some swimming faster than others.

There is a lot of time to think. When you are young you have few problems and little to think about other than how long the swim will be... Hour after hour you kick, stroke, glide. There were no pussy plastic bottles of water in those days. I don't remember getting any food or water all day..... Kick, stroke, glide, head and body mostly submerged, leaving no wake.... On and on... and on.... How does a body do this for 8 or more hours???

There is also the ocean current that lengthens your swim. Hmm... life's currents??

After 4 hours or so, in sight of nothing but a bottom maybe 100 feet down and the sun overhead... looking down in this amazingly clear water there appeared hundreds of Hammerhead sharks. Thankfully the movie Jaws was not due out for another 11 years... LOL. So I had no fear, just curiosity. And for whatever reason the Hammerheads stayed below minding their business... LOL. Feeding frenzy??? And we just swam on...... and on... into the late afternoon. All day.

Finally we saw land and kept on ... Adjusting course to get us into the harbor and to the steel pier. My legs were rubber. The last mile mostly using my arms. In those days our Duck Foot swim fins were heavy and rigid.

We reached the pier. There was this narrow, rusty rung ladder going 30 feet up to the top. Taking our fins off, my swim buddy and I climbed what seemed to be the longest ladder in the world.

On top was a loving instructor who said since we were the first pair in, we would get to squat and duck walk until all the rest of the pairs were in. Calves cramped like *"boing!"*... S.O.B...

How do you love and hate instructors at the same time?

Just training for life....

We slept well.

Gemini 7

The CO called me in and informed me I was to be our OIC of the Gemini 6/7 Recovery Swimmer Detachment.

Being young and stupid I thought going back to St. Thomas with my platoon would be more fun. You know diving, submarine lockouts, explosives, and girls. I had a motorcycle there.

Reluctantly I said "Yes sir", saluted and started a fabulous journey and honor.

We trained for months with NASA and a steel practice Gemini spacecraft; in the harbor of Little Creek and aboard the aircraft carrier USS Wasp. We refined the flotation collar configuration and water entry flyby from the helicopters. It seemed routine to us. Just another job.

But what is routine in the military? Each of us has our own experience and definition; which is usually wrong when you are in your 20's....

The Wasp sailed to the waters off Bermuda where we practiced jumping out of the helos with scuba tanks and entering the water safely. Altitude control and wave time was critical. As was dropping at the right intervals to get the flotation collar deployed and inflated around the spacecraft. We all got good at it. Looking back that was a lot of fun… with training most danger can be minimized.

Gemini 7 was to be the longest duration flight to date, 2 weeks. Gemini 6 was to be the first rendezvous in space, going up and coming back in 2 days. The Gemini spacecraft held only 2 astronauts and was small, hot, and cramped. No physical activity. The main concern was how the cardiovascular system would degrade during two weeks of weightlessness in orbit.

I chose my assistant detachment officer, Ensign Dennis Bowman, to be in charge of the primary helo for Gemini 6.

Gemini 7 was launched on December 4, 1965. Gemini 6 rendezvoused and returned in two days. Gemini 7 returned on December 18.

Splashdown. My 2 swimmers and I entered the water. I quickly swam to the capsule and plugged in a red phone. Were Bormann and Lovell okay? Yep. I wished them a "Merry Christmas", unplugged and assisted in the deployment of the flotation collar and egress raft.

All went without a hitch. They were hoisted up into the helo, and we were left alone with our baby, the Gemini 7 spacecraft. I remember the USS Wasp finally appearing on the horizon with sadness. Our day was over.

Lines were shot as we were pulled alongside, dwarfed by the enormity of the carrier and being just a swimmer in the water.

The spacecraft was hoisted aboard as we climbed the long wire ladder to the deck.

I walked back to my stateroom, wetsuit dripping.

My Gemini swim buddies are now all gone....

God Bless America.

AS-201

The CO called me in and informed me I was to be our OIC of the AS-201 Recovery Swimmer Detachment.

Sound familiar?

Now this time I really wanted to get back to St, Thomas.... And this was going to just be an unmanned new Apollo spacecraft. Well, of course, a manned Apollo would ultimately go to the moon, but I was 25, alive, and still stupid. You know, young and being able to do anything you tell your body to do. A wonderful time of life.

AS-201 meant Apollo, a 3-man spacecraft, on top of the mightiest of all rockets, the Saturn 1B. Now this was a big deal too.

After landing in the ocean this spacecraft would float upright on its heatshield, making for a large sail area, meaning a wind could blow it through the water faster than we could swim. Nice new variable.

We practiced in Little Creek in the freezing January waters, modifying the flotation collar. In February we boarded with all our equipment aboard the aircraft carrier USS Boxer for the long cruise to the South Atlantic.... To an obscure island called Ascension Island. Go google it. Half way between Brazil and Africa. Nowhere.

"The Apollo program is the father of Orion."

We would stay in shape running miles on the flight deck. Kinda cool as no one else could. There was a goofy equator crossing ceremony. Tradition is tradition.

This was going to be the first test of a fully operational Apollo spacecraft. It was a suborbital test of its integrity in space and re-entry.

I remember a special night practice. In case nothing went right and it landed far away. So there we are in a helicopter in the pitch black downwind of the boilerplate spacecraft. 5,000 miles from Brazil and Africa and the water is 3,000 feet deep.

The helicopter's altimeter ceases to be accurate below 50 feet and we have to talk it down to the surface 10-12 feet so I can jump. It was dark and depth perception bad. Jump too high in a trough and you are in trouble. Coaxed down, fingers crossed, I jumped with tank on back.

I was downwind of the module and had a parachute in a bag dropped ahead of me that I was to swim to the module. Then attach so it could act as a sea anchor and slow the spacecraft enough so we could work with it.

There was no communications with the pilot and they were hovering too close with their lights and prop wash limiting my ability to work. Finally they backed off. There was a wind and the module was moving. I attached the sea anchor and pulled the canopy from the bag, but it just floated along, not filling and nor slowing anything down. I'm 10 feet underwater and it is almost black.

I see something out of the corner of my eye. Holy shit, this was one big mother shark. And I am all F….n alone in the middle of absolutely nowhere and no means to communicate with anyone. Go to the surface and it eats me.

So I quickly decide with precision to be crafty and pull the white parachute canopy all around me. Ha, I outfoxed death! As I observed the parachute canopy bag float by….. well, it kinda looked like a shark, you know a Great White. No one could see what happened so I got off scot free.

On February 26, 1966 after a 37 minute flight from Florida we recovered AS-201.

Only recently did I realize that I was the first person to ever touch an Apollo Spacecraft to return from space.

The Apollo program is the father of the Orion Mars program.

To some day take man to Mars.

God Bless America.

The Grenade

Mike Monsoor was Navy SEAL.

Mike won the Congressional Medal of Honor.

Mike is dead.

You see… He, fellow Seals, and Iraqi soldiers they were working with were on a rooftop on September 29, 2006 in Ramadi. A live grenade appeared. All would be killed or injured. Mike reached deep inside where Truth lives and dove on the grenade. His defining choice.

God bless you Mike.

Teammates were saved.

They hammered their Tridents onto his coffin.

Most of us are never asked to make that kind of decision. Really? I am not so sure. Mike defined choice. He defined what serving is. He chose to serve his buddies with his life. He said, "There is black and there is white." He said, "There is evil and there is good."

You want to debate??? Then go in the corner and hang an idiot sign around your neck. I'll bring you some animal crackers in an hour.

"Mike won the Congressional Medal of Honor."

OK, I feel better. But this is all too important to dismiss quickly. Mike became THE role model of role models in that instant. The same can be said of Michael Murphy and unseen others…

Every moment of every day something bad happens. Every bad thing is witnessed by someone. Every bad event is initiated by some bad choice. A person is responsible for that choice. Think about it. People make evil possible. A grenade is a really big bad thing. But… what about the small grenades of life? Does a lie ever have the potential to cause great harm? Does a lie actually damage the liar, much less the person who is lied about?

As we all know there are a bunch of sins I could list and there are the famous Commandments I could refer to in helping to define things that cause harm.

And… How about a simple swear word that begins with an "F" uttered in earshot of a young child? The first time that word would ever be heard. Is there any way to know what seed may be planted…of an innocence lost? Would you want that responsibility? Was this not a potential grenade to this young life? Maybe???

Who is to know? In fact every wrong thing we witness in the course of a day is a potential grenade to someone. Maybe we should look at life through Mike's eyes and react with Holy instinct and attack? Or do we out of habit and political correctness look away and busy ourselves with something else... Some work task that is more important? Or maybe turn up the volume of our headphones and sing "Chasing Pavements" along with Adele...???

I want us to give Mike the Medal of Value. That medal can only be created and given to him by our actions. "Here Mike. I did this because of you. Your example made me do it."

I want to be true to my inner core and see more mini-grenades in the course of each day and right them. Stop them in their tracks. To tell the person who perpetrates that they are doing wrong. I want to do it out loud so others hear.

Too bad if feelings are hurt. Too bad.

Take the criticism and be proud.

It matters.

Be humble.

Every sin is a grenade.

Do something about it.

Thank you Mike.

Hooyah.

Lone Survivor

A mother is standing on a hill grasping the hands of her two small children.

Divorce.

Single parenthood.

A tear unseen.

A choice is weighing down on her soul.

There is a path of short term relief with substances and unnamed people. But those two pairs of hands...??? To quit or to go forward into the unknown, putting those tiny palms ahead of her own.

We cannot predict the future but we can shape it with each positive, unselfish step. Faith yielding hope. One step at a time.

On a lonely mountain in Afghanistan 4 Navy SEALs chose not to kill a shepherd and a child. A choice. You may have read the book or seen the movie. Lone Survivor. Intense. They became trapped. A young man stood high atop a rock to try to get his sat phone to send his location. He was killed. LT. Mike Murphy fell there forever. His mother left alone on that hill in spirit with him. One Lone SEAL survived. His name is Marcus. He survived to tell the story with grace.

Young women get pregnant.

They no longer know if they will be left on the hill.

What has become of us?

If a fetus could talk would she have second thoughts, in this, our celebrated 21st century? Would she pack it in?.... "Forget it. I don't want to be a burden to my future mother. Save a lot of money on skinny jeans"....

What is happening to our men? The notion of responsibility and commitment appears to have been terminated with prejudice by questionable social forces. Where does a young man go for direction? We have even found a way to neuter the Boy Scouts.

Make mistakes and all you get is sensitivity therapy and the assignment of blame to someone or something else. Where is the man walking back up that hill to take the hands of his woman and children? Is the hill too steep? Is not the reward obvious? Ride up on your four-wheeler and load them all on board and get back to being a parent....

Where are the leaders and politicians? Isn't this the terminal cancer in our society? Roman Empire all over again?

The noise of politicians accusing and complaining is heard all across our nation…. like two parents in the final throws of divorce? The family, the heart of our country, will inherit this empty void. This is a bullet at close range.

Forget global warming. What about the erosion of motherhood and family and values? What do we value any more?

Can someone in Washington give us some values to value? The Constitution is a good place to look. Why did they bother to write it? And agree?

There is also The Book….

Mike's mother cried that day.

Her son was cool.

He cared more for others.

He never got to have any children.

He still died for them.

2,000 years ago a mother named Mary cried.

Memorial Day

Our flag flies at half-mast.

Our flag.

The flag of the United States of America.

Our flag allows us to debate, criticize and protest as much as we want without fear of persecution. It is amazing how much we love to criticize. Media and social networks are ablaze with condemnation. Oh yes, there are stories of good works… but those headlines are stolen by scandal and criticism, however intellectually cloaked.

Freedom of speech is the greatest threat to freedom throughout the world. What we take for granted is feared elsewhere. We are under attack by radical this and that because freedom of speech will destroy their power. Terrorism is applied to those willing to speak. Sad.

Graves freshly dug and graves of 1776 abound with sacrifice and tragedy. Tragedy is the life that was lost for no reason. Those of us who live have to make those losses have meaning by how we conduct ourselves. The veteran is not that outspoken for he knows the horror of war and the value of freedom. We all too often mock and abuse it.

"*Thank you for your service.*"

It is time we bring dignity and respect back into our lexicon. Yes ma'am, yes sir. Yes Mother, yes Father.

Our hypocrisy is so much more visible these days. Look at the killings and the excesses of our behavior. Look at pictures of crowds these days and how fat we have become. We mock those who differ. We mock religion… Respect for elders??? Forget it. They are just old people who are not relevant.

On Memorial Day we are bowing our heads to our cell phones and not to those who made this land of cell phones free.

The tear in the eye of the Veteran reflects the light of Truth.

It knows evil and it knows folly.

Thank you for your service.

God Bless America.

Leaving There

Discharged

The CO calls you to his office to wish you well and thank you for your service.

Ok, that is what a good CO will do.

Otherwise you sign a form, grab your papers and duffle bag and walk out.

You may have been in for 4 or 20. But you were "in" and don't belittle it. Compared to all the men who were never "in" you stand taller. But false modesty keeps most from feeling tall.

Your last beer with your buddies. A camaraderie never again to be known. In combat there is a rarified bond, unspoken and giant. Even so, all are part of the larger brotherhood of the military.

No teammate…. nobody left behind. The moral code of mortals. The sea of life needs swim buddies. Think about it. Wives and family for sure, but a fellow soldier is mandatory.

You get off the bus or plane and are greeted by joyous friends and family. Others walk silently through the terminal to some taxi or car or train. Hearing "thanks for your service" was never heard by many. It hurt. The pain silently lingers.

"Ok, that is what a good CO will do."

Home and then the civilian routine begins. No danger, no buddies to joke with. No person who has seen what you have. It is almost like pretending it never did happen.

Employers without military background shuffle you around, not knowing that you know more than they do. The pride when in service is muted now. Who do you go work for? New forms to fill out. Now with more fine print than Einstein could decipher. Learn a trade? Work for a corporation? New politics that you never see coming.

Nights are fun until the morning, when dreams steal courage.

You have no choice but to go it alone.

Just do it.

Climb out of the trench and run forward.

And find a buddy.

God Bless America.

Not Lucky

What does luck have to do with it?

Ask the person who was not lucky.

And those who were lucky sure know it.

Luck could be a thumb blown off. Luck could be a flesh wound. Luck could be a broken limb. That's lucky.

Today we get to see videos of the indifferent brutality of war and false theology. Blood and bodies, children and women, and soldiers. Not Lucky.

The Veterans Administration is full of not lucky long term patients. Rehabs for every possible injury. But not rehabs for the soul. The most private part of the human being. Drugs and surgery can't get there.

You don't know you are not lucky until the doubts invade from dreams unseen. Dreams of sights seen, of deeds done that impact one's most private feelings. You can tell a psychiatrist but you can feel no bond… maybe empathy… but you don't want that. You want answers. A way out of being "not lucky".

I have not seen some of what you have. Your family has not. But talk to

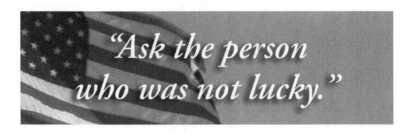
"Ask the person who was not lucky."

me, one of your brothers, and I will listen and understand and not let you down.

We all need a "Mr. Not Lucky" to whip us back into shape. Call it a spiritual boot camp.

There is no reason at all to not feel lucky.

To turn your scars and knowledge into something that serves someone else.

Spiritual casualties need spiritual casualties that have not quit, who have found a path to real value.

All built with the blocks of pain and insecurity into foundations of respect.

Every veteran can always be a hero to someone.

"My eyes have seen the glory….."

God Bless America.

The Men

Damn it, it's the men.

Not my civilian men friends.

Damn it, it's the men.

The men I *served* with.

Whether it was B-17 crews in the 8th Air Force or the sailors on DDG 112 you *served* with them. The men.

Whether in danger or not, the intimacy of meals together, joking together, drilling and training together, or venturing into the night together... it is always about the men.

Always concerned about the other's safety. Always in an unbeknownst-to-you mindset of service. You had people to serve. It's not the same in a civilian business. It is not the same in a bureaucracy where selfish is the norm.

To serve is unselfish.

To serve another person is beautiful.

To serve your country is beautiful.

"*Damn it, it's the men.*"

To serve a fellow veteran is beautiful.

Damn it, it's the men.

God Bless America.

The Hand

"Take my hand".

You will say to any boy as you cross the stream.

For you know that a slip can mean a fall.

And a big rock could be bad for a small head.

Take my hand is a reflexive instinct that every human has. Why?... Why is helping a child or adult so natural. After you cross the stream and share a quick glance of bonded pride and gratitude.

Look at your hand and think of all it has done and can do. It can write, paint, drive, caress, give assurance, and shoot. The list is endless. It can bleed and break too.

When the veteran arrives home, tons of hands are offered. Shakes, fist bumps, high fives. Hands in action doing their thing. Then… there is silence. All move on to their worlds and the veteran's hand is unclasped. This is the critical moment. Hands need hands… and special ones for special needs. Every veteran needs the hand of another veteran every day just as it was in the service. But they don't know it. Fraternity for eternity.

"All hands man your battle stations."

The firm grip, eye into eye. Hands and veterans bond. No hand should be without that hand.

Service you have been in. When out, your service is then to locate a fellow veteran that needs a friend. Most all need one though they pretend otherwise.

Talk about work, talk about wives, talk about fears, talk about lives. Today you can text a vet. Hey, that sounds like a website??? "TEXT-A-VET". The point is for Vets to stay connected.

Hands off? Bull shit. Hands on guys. We have to be more interactive and proud.

Do you know all the school kids out there who need to see you raise your hand and salute the flag?

Who need to see you salute them in a classroom?

Who need to know that being in service still means something to you??

On ships a sailor is called a Hand.

"All hands man your battle stations".

And find the hand to hold.

The VA

Yep, Norfolk is in Virginia.

VA is a state.

VA is a state of being.

Because if you are in the VA, like as a patient, you can get fantastic medical care. Let's give her the benefit of the doubt as there must be a lot of ex-military on staff. A lot of people who really care.

We have all heard about the long lines and stupidly debasing long waits. Hello, it is a bureaucracy. Layers and layers of decision making and reports and opportunities to delay.

Field hospitals work better. Right?

Can't you and I swim together through the maze? Share the discouragement. Share the night dreams. Two can make more noise than one. Doctors that are too busy can prescribe pills. Most are psychological overdose. You know what a foreign substance in your bloodstream can do. The complex chemistry of the body is beyond imagination.

The VA psychologist can only try… and there has to be a mutual simpatico and trust and availability… to find answers.

Blame the VA?? Blame us. We vote and we trust. But have we lost our way on the way to the VA?

Let's call the VA the Veteran's Alternative. Find answers beyond the VA. The VA is needed for the crippled. The Veteran's Alternative has to do with love. Love of self, love of Truth, and love of God.

Sorry Charlie, no God in the equation and you increase your odds of failure. Prayer with your buddy once a day is mandatory…any form of prayer…. Just a text saying "Hi" and "God Bless You". Prayer is not just kneeling in a church. It is more than that. Much more. For in prayer you look for Truth and the right option. How to serve someone. How to help. That is prayer.

We have to ask more of ourselves and less of the VA.

We have to slim down our bureaucracies before they sludge to a halt.

This can only be done with love and prayer.

The Choice

Some say we have no choice.

Some say we do.

Our choice is to determine which "some" we are.

Hard to remember back when you made your very first choice.

As a kid you really knew nothing though you thought you did. Same goes for your teens and beyond. Sometimes choices are stupid, or just uninformed. Can you blame someone for a bad choice if they didn't know the facts or were being guided by the wrong person or perspective or feelings?

Is a good choice from luck or true decision-making that *you* made?

We make choices when we vote… don't we? Are we fully informed or being swayed by the assertions of media?

The veteran made a big choice to enlist and a big choice to serve however long. The veteran makes a big choice when he gets out. About job and family and friends and past. The best choice is to just keep moving, to increase your odds of landing right. That's a big choice not to quit on finding choices.

"You can't go back to the war zone or in time."

What helps one make the right choice? First of all you have to know the Truth. With Truth you can assess the facts and options clearly. Certainly more clearly than without Truth. Think about it.

Most know in their gut what the truth is. At the center of self lies the truth. It is that feeling that comes from your heart. Be honest to it and you will feel no lingering guilt.

Many veterans have to deal with hidden guilt. Guilt that can only be eased with forgiveness. Who can you get to forgive you??

You can't go back to the war zone or in time.

Friends can help.

But you have to stand tall and forgive yourself.

If you look up into the sky you are forgiven.

But you have to kneel first.

The Choice.

Withdraw

To withdraw is the most difficult decision facing a commanding officer of unit, platoon, brigade, and theatre.

Do you go forward against all logic and risk lives that look to you for the right decision?

History tells us of every form of consequence, good or bad.

McArthur did in the Philippines and came back to victory.

Can't we withdraw and come back? Of course… life offers options and especially for those with faith it offers hope.

The veteran returns from wherever circumstance sent him…like the shuffling of a deck. Four hearts or four spades….. He returns with a smile on his face, but with memory subdued. The memory is asked to withdraw. Bones can be broken and heal. So can a memory with the corrected perspective. Drugs can't correct. The love and insight of a buddy can. A fellow veteran.

Analysis fills reports, but not the soul.

We all hide memories. Sins against friend or Truth. We hide them. They haunt moments unexpected.

"Can't we withdraw and come back?"

Withdrawal is mandatory until you are strong enough to confront your own Truths. Say "I am done with guilt memories". "Time out". Then pray and pray and pray and pray. Didn't you once pray when you were shoulder to shoulder with danger? Or when your wife or mother was delivering a child with difficulty?

Withdrawal does not have to be negative. It is only negative when it is withdrawal into substance abuse or Truth denial. Drugs and alcohol are dead-end choices… the ultimate withdrawal and loss.

Not living honestly and in Truth is just as dangerous.

Withdrawing into God and is an option that has to be considered.

Spend a little time with someone who has.

This is not to say that hypocrisy will not exist.

Just ignore it.

Everyone knows Truth if they will admit it.

Withdraw only to return.

Otherwise withdrawal is nonsense.

Suicide

If I had committed suicide I could not have written this.

If I had committed suicide I would have created even worse dreams for all of the family, friends, and children I would be hurting.

I would have sent a message that there are no options other than suicide.

If the VA can't help, much less get me an appointment, then I will fight for my right to serve again. There has to be a field hospital of Hope. Like the USS HOPE (AH-7).

Damn it, I will not let you quit on life.

If you have seen the real horror movie, and tasted the real shit then you have the knowledge to serve again. Combat in civilian life is for the Truth, honesty, and caring. A fabulous battle to be waged in your 10 foot radius. Wherever you are you have opportunities to stand for something. In most every moment. Why can't pride replace guilt and pain? Isn't one kid's smile worth waiting for, much less creating?

Why do we put ourselves down so much? Doubt so much? Have we not seen extremely poor people sing in church?

What is the message of a gun to your head? Not that you don't count, but that everyone else does not count. A smack in their face as they cry at your grave. You left them with no hope. And they still love you.

"Suicide is a lie."

Let's say No to self, and be reborn into a man with the mission to make life better for others. If we have fought for them, if we have agonized for them...then we have to complete our mission and not quit.

Having lost a fellow soldier or having killed an innocent is a private burden. But quitting on Truth is not acceptable nor what they want.

What if you knew real Love and looked above...?

If you got out of self, and no matter how broken, and you cared for others during your pain, you would emerge as a greater hero than you ever could have imagined.

Have you seen men who have made it through AA? They have smiles and pride. What the hell from? Just by sharing their Truth and listening to others who have known greater pain? Amazing. They just call it "higher power".

Give me a break.

It is God.

It is Jesus.

Die in His arms.

Suicide is a lie.

DDG 112 Hooyah

When something bad happens we all stop a moment and react.

Like when the earthquake was felt this week.

My wife and her girlfriends got scared and confused. Natural.

But when I talked with my buddies their immediate reaction was as mine… "What do I do right now?" "Who needs to be protected and how?" That is okay; that is the way life has been created. Nurture and protect; our primal roles.

Some things are meant to be natural, left alone… as they are part of nature, our nature. We are born. We die. This is natural. This is accepted. We can leave it alone or we can try to alter. Man can dignify it or man can desecrate it. Good or evil. Selflessness or selfishness.

Peoples' lives can pay tribute to life or they can cheapify it… (New word, cheapify, or cheapen… I like cheapify…) Hitler made life cheap. Our soldiers ennobled it. Love makes life beautiful. Addictions make life ugly.

When something bad happens something has to be done about it. "Done" does not mean debate! "Done" does not mean politicize. Nike says "Just do it". So do I. Attack bad, don't dither it. Inaction is the very worst form of action.

> ## "What one does for others is the measure of one's life."

We have a military because history has taught some of us that bad things have and do happen. Diplomacy works often but not always. There are people who lie no matter how much we give and try to understand.

We may even have to act without anyone else's permission. While all the public posturing is going on, there is not a day when someone in the military is not somewhere operating covertly, much less an innocent drone hovering for the next "perfect" kill. Our Special Operations and Intelligence communities have their lives on the line daily to allow us the delicacy of thinking peace is at hand, that we are safe. If something goes bad… our good guys are sent to situations you would cringe at, or withdraw into psychosis.

DDG 112 USS Michael Murphy was christened in Bath, Maine on May 7, 2011. I was honored to be there. I knew Mike at the start of his journey. Many of you may have read the book "Lone Survivor." Mike was a SEAL officer who was sent really deep into the bad. On a lonely mountain in Afghanistan he and two of his men were killed by the Taliban. One miraculously escaped and lived to tell the story. Mike took his final bullets standing with his radio so he could get clear transmission.

What one does for others is the measure of one's life. You can take bullets on a hill or you can take criticism in your home. It does not matter as long as you are standing up for good. Calling a spade a spade… calling bad for what it is, period.

We have become passive in our beliefs. We have become afraid to say out loud what we really feel, what our heart tells us. We don't listen to our heart because it may be politically incorrect. Proactive or passive about life? About values? About evil?

A year later in NYC, at the end of the commissioning ceremony on the Hudson River a yell by the 1,000 people there of "HOOYAH MIKE!" resonated in the piers and hearts nearby.

When I was told this, my stomach turned and my heart became heavy… and tears… because the word "Hooyah" is so unique to a very special community in Naval Special Warfare to which I once belonged.

Godspeed & Hooyah Mike.

Eyes On

"There they are."

"Adjusting the windage."

"Are you sure?"

"Send it."

Sometimes decisions just have to be made. Every adjustment to clarify the Truth is essential. What more can you do? Not to face evil with immediate resolve risks death of the more innocent.

There are moral moments where an immediate decision is necessary. But the allure of pleasure clouds the process and the windage cannot be set and the target of safety is missed. We go through life seeking the pleasure of acceptance and material comfort and sensual sojourns…. Only to find we have missed the target, our own self-fulfillment.

Read the papers where so many celebrities and wealthy and poor are unhappy when they reach their later years….. Fake smiles and tales of happiness hide their Truths. For happiness only comes when you help someone else. Try it. It is seductive too.

The eyes have it.

Most everything is brought to us first through our eyes. Eyeglass frame

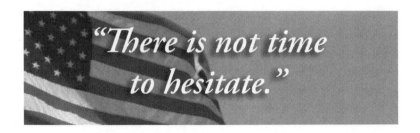

"There is not time
to hesitate."

designs are in the millions. All to make your eyes look more fashionably better and to see better. Transition lenses, trifocal, mirrored, sport…. Designers are making fortunes just off designing frames with their names on them. On your eyes!!!

Optometrists are working overtime to help us see.

I sure love what my eyes can see. When I was young it was all about activities and pleasure.

Then I started to see with my heart and things changed, really changed… I could see how happy people were, but also how sad. I can see subtle pain with the slightest glance. Opportunities to reach out and reassure are now in abundance.

Amazing. Amazing. All around me. I never knew it was there all the time.

It hurts seeing other's pain.

But it is an honor.

Humbles one.

But there is no time to waste.

There is not time to hesitate.

"Send it."

VVV

HHH is a professional wrestler.

BBB is the Better Business Bureau.

AAA helps you with your cars.

Triple X was a movie with Vin Diesel and is a symbol of evil.

Triple Crown is the ultimate in horse racing.

Brand new is Triple V, VVV, Veterans Visiting Veterans. An idea I hatched for my Vets in church to be more active.

We established a relationship with an assisted living facility and met with Veterans there. My concept is for each one of us to meet once a week for one hour only with the same veteran. Each having their own different individual. Just for conversation and companionship, nothing religious.

You well know that we Veterans have our own language and are rather private about our experience.

There are veteran's associations out there, but none come close to this....

But when away from beer and alcohol in a private setting the sharing can be phenomenal.

"*Veterans Visiting Veterans.*"

When coaxed and assured that you are sincere, the stories flow from WWII to Korea to Vietnam to the present. Whether a clerk or a pilot, all served their country with pride. How they enlisted is a blast and humbling.

Telling the truth to fellow comrades is like a breath of fresh air. The humor is still there. Special.

A new friend, old Steve, has taken ownership of this ministry for me. His time in the service in Vietnam was rough. Hard to forget. But he has risen above the pain.

All who served are privately proud but quiet. Only 5% enlist today, the rest have no clue.

Our veterans all believe in a universal draft too.

VVV. Veterans Visiting Veterans.

Triple your pleasure……

Kinda cool? Eh??

One Veteran

One veteran.

That's all it takes to make a difference.

Just one.

Saluting his flag seen by one pair of eyes.

There is a truth in that.

It is the veteran who gives the salute meaning.

Whether he endured pain or not.

The uniform inspires.

It exists to protect good.

One veteran adopting one veteran.

Let the clerks decide how to process.

But the veteran waits not for approval.

For in his heart he knows what to die for.

"That's all it takes to make a difference."

He knows how to carry the weight of another veteran.

Until he drops.

God Bless America.

Eulogy

When it comes time for the eulogy… who is going to say what?

I wonder if that person is going to say the right thing.

I hope they don't ask that other person to say something.

Will my wife be crying?

Will my daughters say the right thing?

Sure glad I will be busy elsewhere.

Spare me the video.

I hope it will be in a real church with real pews, not on the beach and ocean which I loved a lot too… After my stents and bypass surgery I got to live well beyond my Dad's years…if this is an accomplishment. Did the world benefit??? Or at least my world??? Or was I just a consumer? Of time…?

I have heard others' eulogies and they left me kind of cold (pun intended). They talked about the love of family, vacations enjoyed, hobbies and sports. Business career and other accomplishments… even generosities… and "he loved this and that" ad nauseam. For the most part they sound a lot alike.

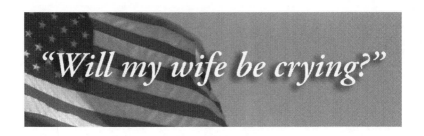

"Will my wife be crying?"

The pursuit of titles, money, or recognition traps so many on a street to nowhere. At eulogies you hear litanies of the material and passing allusions to compassion. Religion??? Gotta be careful…

What were you meant to be?… rings in my ears… did you achieve it? Did you figure out what life is all about and make sure you fulfilled your promise?

Somehow I feel that journeys end too soon. Life just can't be about the last years of golf… or finding a good church, and watching your pennies. Who were you meant to be? Or who could you have been???

Did you stand up for what is right? Did you define good and fight evil?

I have an intriguing résumé: Yale, Navy SEALs (UDT), great daughters, great travels, finishing with my wife's incredible business story. Who would have believed???

Yet all this would be meaningless if someone did not mention Jesus and my extraordinary involvement with Him. Only the few know I was His rogue apostle chosen to lead others outside of the walls that kept them from being splendid spiritual butterflies.

That is the story worth telling.

Joys to share with others that transcend all other relationships.

Boy, was I blessed to have the perseverance to not quit on God.

If I were alive during my eulogy I would shout out loud, "Don't quit on Jesus!"…

What you can't yet see is just around the corner if you keep moving forward.

Godspeed brothers,

Hooyah.

Last Battle

Call To Arms

Veterans know this country better than anyone.

From Private to General, from clerk to pilot.

Any kind of service to country makes you a shareholder.

A veteran has known sacrifice however mundane.

He sees his country adrift.

From values to leadership.

A call to arms.

Arms that now must vote and create votes.

The battlefield is politics.

The battlefield is bureaucracy.

Purple hearts to those who stand in the line of fire and speak from the heart, taking criticism from all who find the status quo easier. Men are born in the services to serve. They give up identity in uniformity. The military is united under rule, rank, and patriotism.

It is time to serve again.

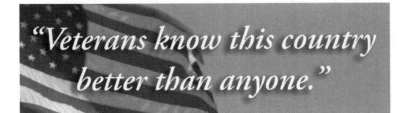

"Veterans know this country better than anyone."

To not care about opinion but that which lies in the heart.

"Forward…. March !!"

God Bless America

Valuedise Lost

Our nation was founded on core values.

Our first universities had mission statements that all revolved around traditional principles of honesty, truth, integrity, and doing "good".

All came from protesting backgrounds.

They were all Protestants.

They had to fight for their values. They were the cornerstones of this country. Immigrants learned English. We were rich with diversity. In the homes were strict rules.

The United States of America used to be united. Looking at TV today we appear to be the most divided nation on earth. Republicans hate Democrats. Democrats hate Republicans. Atheists hate religions, especially one. A nation of incessant protesting. Minorities now ruling majorities. Go figure???

You are not allowed to discipline a child unless it falls under legal guidelines. LOL.

At the dinner table heads are down into texting cellphones. Parents, now single, can't fight the "whatever" attitude that confronts them every

"All came from protesting backgrounds."

moment. The downside of social networking is false ego affirmation that comes from the inherent instant gratification that your friends like you.

Where do kids find values; some ethical or moral compass to go by????

Not looking down.

We used to have Boy Scouts. Parents could say no. Parents could discipline.

Single Moms can't do it all. The family is suffocating. Schools run away from values leadership.

Islam is not allowed criticism. God is.

Churches are mocked. It is not politically correct to acknowledge God inscribed into our municipal buildings and monuments. Upside down??

Paradise Lost.

It seems to me that the last bastion of order is the military. You are told what you can and cannot do. You learn to respect the flag and country. You learn the value of authority and organization. You learn comradery and trust. You learn about teamwork. I know. Have you been through Hell Week? Ok, I went through it and learned to swear and to talk trash along the way… It's how I survived running 4 miles on a beach in the sleet.

(PS. Walk out of the room now. I believe in the universal draft.)

Let's declare war on the attack on values.

Let's defend our kids and families.

Let's stop our paranoid worrying about the feelings of others and fight for what is "righter".

We are fools if we don't think our Rome will burn.

Let's be value driven and stop all the hogwash infighting.

Let's make it Team America once again.

If not, the new Pearl Harbor and the next 9/11 will bury us.

Valuedise Lost.

Want Better?

Everybody wants better.

Better cars.

Better homes.

Better money.

Better hair.

We could list a hundred things we want better without effort.

But wanting better is not getting better. Better only happens if you make it happen. It is also best to list wants in order of importance. Whatever you want the most cannot be distracted by the # 2 want. You have to make real headway on #1 to gain the confidence to juggle #2. The world has to see you make progress.

Getting better also requires a real commitment. Real effort. Real pain.

Doubt will attack you. Not quitting will see you through.

Better looks so rosy and perfect and easy to attain. But it ain't. Most people want easy better. And then they don't get any "better"….

Look at yourself in the mirror. Do you see better?? Do you see want? Real want?

"Our forefathers wanted better."

Everybody can get 'better" if they really work at it. But to get "better" you have to be totally honest for starters, especially with yourself. Then you have to learn respect and manners. Nothing is going to change until you do. Then you have to learn to help others. To get off your "My Horse" and mount the "Your Horse".

Want to make America better again?

Start with the same approach. Be honest and caring.

Demand more from your politicians. Learn more about issues. Make your vote reflect how you feel. Better requires a price from all of us. There will be rough times.

The kid will scream and yell until the "NO" is unwavering. You know it could get nasty, but the easy way is to stand fast. Better is at the end of a tunnel. Don't quit.

Our forefathers wanted better.

They wrote a Constitution.

It was seeded in some 10 Commandments from a long time ago.

Believe in Better.

Quitting is not an option.

Be Strong

To stand up for what is right you have to be strong.

To say no to drugs you have to be strong.

To be a good parent you have to be strong.

To be a good politician you have to be strong.

To be a good citizen you have to be strong.

If you want to be strong you cannot quit.

To be strong you have to have beliefs and believe in them. You have to have values. You cannot wait for them to come to you. You will be uncomfortable but you have to go out and find them. The hard way. On your own. Make mistakes and learn. There is no easy way.

A political campaign is demanding and not comfortable. Questions, travelling, criticism, coffee, beds that are old, schedules, interviews, debates, advice, and a unique aloneness. You gotta be strong or stupid.

You have to stand for something, not sit.

Be the target. And don't flinch when you hear the trigger. There is always something headed your way. Concocted by others in other places. Wanting to test your strength. Not caring. Just looking for weakness. That's life.

Being strong means admitting that there are forces at play to keep you from being strong. Whispers in your ear that are false. Loyalties that aren't.

At the end of the day is our ability to make a decision on the spot that is a decision and not an equivocation.

The power of the word "NO" is when it is steadfast.

People have to know you mean what you say.

We get all caught up in nuances and sentiments.

The dither in media and politics.

Be strong and say NO to this waste of time.

Put yourself on the podium in all the bright lights and stand for something.

Be strong.

Made In America

I was made in America.

I was born in New York.

I was not born in Africa.

I was not born in China.

I was not born in Iran.

I was not born in Mexico.

How come I was not born in Siberia? I don't know. But I'd like to meet the Guy who made the decision and shake His hand.

Why does everyone want to come here? Can't they just leave us alone and solve their own problems. We did ours… Kill off as many of your own as you want. Just leave us alone. We will even provide a little foreign aid for you if you do.

We don't have anything you can't have. Liberty, Justice, and Freedom for all. Just do it. Equality? Human rights? Just words that make such good common sense that everyone understands. Capiche?? So get on with it. Google all you need to structure your country like ours… cast some

"I was made in America."

votes… throw out the bad guys… show women some respect… and be kind to animals… and… you may get even more foreign aid.

Okay America, what the heck is wrong with you? Why are you setting such a bad example to the rest of the world? Great job in managing family without fathers. Great job in legalizing nonsense and pot, metaphorically speaking. Great job in legalizing the legalizing of everything so fine print is your new god. Great job in idol worship at the expense of value worship.

Yet everyone still wants to be 'Born in the USA'… (Thanks Bruce Springsteen). Go figure.

Now we used to give birth to all kinds of products in the USA… But somewhere greed took over and managements in both business and labor started to get paid much more than they should. The rest is history. But everyone wants to become an American citizen. (I hope it is not for the entitlements…)

You know we put a man on the moon. I just bet we could build factories in our homeland and put man back to work. I just bet CEO's could make it happen if their bonuses were at stake. And, by the way, they aren't the bad guys, they are the good guys… if they would reconnect with their values. Their leadership is essential. But they must work alongside the worker, shoulder to shoulder, to build mutual respect.

Ask a veteran how it works.

Ask a Navy SEAL how it works.

Made in America.

Heck, let's make everything in America.

Let's make pride and trust and teamwork more than chalk on a Harvard blackboard.

Poverty

Nobody wants their kids to be poor.

We hold them when nursing so they feel wanted.

We clothe them and feed them and nurture them so they feel secure.

We watch them closely so they acquire feelings and understandings that we hold important.

We know a little of what poverty can be like and we try to shield our children with education and positive activities.

We do our best.

Then they are on their own. The world is never as they expect it. We put them in the raft, then push it out into the current in the river of life to take them where it will.

If the river is in the United States, there is hope. But suppose your river is in Syria, Ethiopia, Yemen, North Korea, Iran, Egypt, Saudi Arabia, or even Russia or China? Try and put yourself in those kid's shoes... if they have any. Maybe you should teach your own kids about the real world beforehand; her history and her poverties.

Socialist states create massive entitlement mindsets and stifle individual potential. More and more people are turning to government to provide

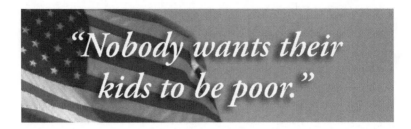

work and easy compensation and insurance. Bureaucracies become inefficient cancers. Reaching critical mass, the majorities feel entitled and the society crumbles. Poverty kills. Entitlement kills.

Our kids are getting spoiled. There is a digital culture that is providing instant gratification. Keyboards are rewarding impulse and that is leading to entitlement. Parents are being expected to continue providing for everything and future security. Entitlement is becoming the arrogance of the ignorant. Woe to us all.

No one is entitled to anything other than liberty, freedom, and fairness. Look at all the countries where this is an illusion and a lie. What are your odds of getting anything if you are black, female, and born in the Middle-East?

Your mother's love is the only brief glimpse of fairness you will ever see.

You are entitled to nothing other than depravation, pain, and uncertainty.

We should send all our kids over there again in a new Peace Corps, but call it Fight Poverty Corps.

Tell it like it is.

Serious Politics

Now this is a funny chapter.

The hard part will be not laughing out loud. LOL

Now it is very easy to criticize politicians these days.

The House, the Senate, the Governor, the secretary of this and that, and even the White House. They have asked for it and they got it. LOL. Except we voted for each one of them. LOL

The world is spinning out of control with the evil in other governments. Yes, Evil with a capital E. We won't admit it. We use words to parse it away. New phrases and categories to make it seem benign. Shit people... Evil is evil! It is not a war on anything else. Evil is created by pride, vanity, selfishness, power, and poverty. Injustice is rampant. The abuse of women and children is astounding. Yet we indulge in meaningless protest and jabbermouthing. We act like we don't want politics to get serious as we rush from one fire to another. Ebola is symbolic of everything that is out of control.

Kids mouthing off. Respect for family and authority going down the drain. Respect for all that we have built in this great nation is under assault. The media observes and turns it into entertainment.

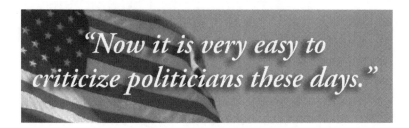

"Now it is very easy to criticize politicians these days."

Who is laughing out loud? We had better answer that question. What red line has been crossed? Do we have any red lines?? Do you personally have any red lines anymore? What are they? Are you willing to tell someone else and risk criticism? Do we stand for anything?

Serious politics requires serious politicians.

Serious politics requires serious voters.

This is no laughing matter.

Constitution

My constitution is fine.

I am feeling better.

I am a lot older and had heart surgeries but have really changed my diet and I am exercising again.

I feel surprisingly good.

My constitution is back to normal.

Less alcohol, less white bread, less sugar, less eating….. It works.

What we do with ourselves determines our constitution. How we feel about ourselves. If we remain centered in self, we just will not feel good, even though we pretend otherwise. The best and only food for a good constitution is helping others. Each time you try it you will like it.

There are rules if you want to feel good, if you want to have a good constitution. There has to be some church with some pastor who can resonate with you. Try to find him or her. Don't quit. Don't prejudge. It is its own arrogance.

We need healthy and wise citizens to lead us out of our bickering

"My constitution is fine."

political chaos. We need to stop amending our Constitution so it is more politically correct. The spirit of our Constitution has served us more than well. There is nothing wrong with it. Fine print cures nothing.

Our founding fathers used Christian principle as our cornerstones. History affirms its solidity. We have fought so many wars trying to protect the inalienable rights we hold dear. We are going to have to fight again soon. Islam rejects our values and culture. We have to stop debating otherwise. We can live in peace only when Islam chooses to champion peaceful co-existence.

Their leaders have to show some leadership or the price paid will be high.

Where is their Constitution that affirms inalienable rights?

And respect and freedom for women.

What century do they live in?

We need peace and honesty for all.

Evil must be recognized for what it is.

EVIL.

Say Something

"Say something, I'm giving up on you" are lyrics from a wonderful song of the day.

What does it mean beyond a broken love relationship?

Beyond the song?

Our nation was founded on principles that were hammered out by a group of men with a united vision. They didn't want things to be the way they were. They envisioned freedom and justice for ALL. Absolutely no laughing matter.

In our new journey into political correctness, polarization, negative advertising, and accusation it seems as if everyone is guilty until proven otherwise….. yes… right here in the USA.!

The polls indicate our intentions before any vote.

Smear and fear are on the front pages.

One may be afraid to go to the polls to vote for fear of violence??

What is terrorism but the creation of fear??

Radical this or radical that, murder is murder and deserves no leniency.

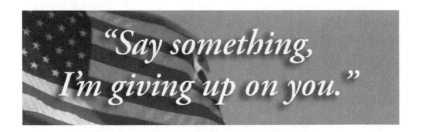

"Say something,
I'm giving up on you."

When did we become so sensitive that the murderer can have his own radio show from prison??

Is our Constitution saying "Say something, I'm giving up on you"??

Entitled

Now this is going to be the most difficult chapter to write and the most difficult to understand.

You are entitled to a full explanation of what is due to you.

You should never accept anything less than the truth.

That is your right.

From your parents came the decision to gamble with your existence and after that magical moment you were entitled to your first breath. No one has a right to take that away from you. Tell me if that is what you think. But more importantly are you entitled to love?? I think so. What is the point of life if you aren't entitled to love? Think about it. You have to make some decisions at some time in your life. Think about it.

What has value and what does not? I think we think we are entitled to value. Well, we get to make choices on what to buy based on value. We are entitled to that right. The right to purchase what we want, when we want… right? We are entitled to love when we want? Right? Are we entitled to good behavior on the part of others??? Are others entitled for us to act responsibly and with compassion? A lot to think about.

Are bad guys entitled to justice? Are nice Democrats and nice

Republicans entitled to a country safe from harm and debt? If I am rich am I entitled to respect? If I am poor am I entitled to respect and compassion? If my neighborhood has been invaded by crack dealers am I entitled to a way out? Maybe….

Years ago, before you were born, the concept of entitlement was seldom heard as everyone was so busy working and helping one another. Hands were held and hope was the force that launched our amazing nation. Immigrants from England, Ireland, France, Italy, and Spain came on boats of all sorts. Nothing like today. Conditions were marginal to say the least. Ellis Island was full of our fathers looking like deer in the headlights. They expected nothing. The only thing they felt entitled to was freedom from discrimination. That is what the United States offered to those wishing to gamble their pasts and plights. Only hope. Entitled to hope.

WWI and WWII were fought as we felt we were entitled to freedom. Look at every cemetery from Normandy to Arlington. Seas of white freedom crosses. They were entitled to their crosses. They are entitled to our prayers and gratitude.

Whatever happened to our great country? Yes, there has been selfishness and greed and crime that created the need for more laws. Laws require millions of people to obey and as many to interpret and enforce. Bureaucracies of enforcement and code grew under the radar screen.

Paperwork justifying paperwork became a cancer. Insurances that insured everything metastasized. Groups of interest unified and these unions became monoliths where members now became managed rather than served. Somewhere entitlement became the vision of this new culture. Be lucky enough to work for a public bureaucracy and you have pensions and insurances second to none. Special interests protecting special interests. The politics of survival has become entitlement.

In this new world order of entitlement our young are just expecting more. Our culture is one of instant gratification. Media, video, headphones, and the iPhone have become the Pavlovian conditioning elements of entitlement. With jobs scarce and debt responsibilities ignored at both the personal and national level, chaos is near.

Kids expect money.

The next generation expects to be paid more for less.

We oblige and nurture the "entitled"….

While they text one another…

I am entitled to breathe, love, and… to be entitled.

If so, then the end is near.

Vote Warrior

Veterans' terms.

No discussion.

Votes count.

Attention on deck people.

Stand at attention neighbors.

This country needs us all right now.

Neutralize and remove redundancy in government.

Get in the face of your congressmen and senators.

Fly your flag in every citizen's face.

Too bad if they don't like it.

Every vote sends a message.

My family and my flag rule.

Values and order are in demand.

God is good.

"Veterans' terms."

Evil is unacceptable.

Vote triggers change.

Stand for something other than feelings.

Stand for Truth.

"Forward.... March !!"

God Bless America

EPILOGUE

I have been blessed with a journey that has not seen the real poverty or felt the real pain of the majority. Please don't judge me by the immaterial, much less the material. I was born in Bronxville, NY in 1940. Grew up in Louisville and St Louis. Graduated from Yale and went into the Navy. I had the great honor of fulfilling my dream to become a Frogman. I graduated from BUDS Class 31E, Basic Underwater Demolition/Seal School. I was an officer in Underwater Demolition Team 21 which became Seal Team 4 in 1984. I had the honor of recovering several spacecraft, including Gemini 6/7 & AS-201, the very first Apollo Spacecraft to go into space. Wow, did I luck out. Then I spent 40 years in women's retail, in various department stores. Even a year at the World Wrestling Federation... go figure?

I have two great daughters and two grandchildren who have just discovered the water and facemasks. My wife has created probably the #1 women's accessory store in the country as evidenced by how much she is copied. Therein I work and report to her... No comment. LOL.

As you can tell by reading between the lines there is a spiritual side to my journey. Kind of covert as I just want to make a difference unseen.

God Bless You All... Happy Trails.

ACKNOWLEDGEMENTS

Writing 1-800-FOR-VETERANS-ONLY just happened while writing 1-800-OH-MY-DONALD. I had to do something for Veterans. I had started a Veterans Visiting Veterans mission in my church and at an assisted living home got some veterans to share. And share they did. Wow... humbling tales of WWII and Korea... all were so anxious to talk. And in a private atmosphere, no beer, the stories poured out. Veterans need veterans. This book is to all who are alone regardless of age. And to those I call again to serve them.

Special thanks to Rev David Ames and Jim Stevens, Roy Bruton, Bill Lord, Sue Stedman, Rob Olson, Juan Hernandez and the many others of my Veteran Ministries in Naples and Kennebunkport. And the Brotherhood.

The chapters average only 300 words. You can open the book anywhere and read the 2-3 pages and get a smile and a frown if you catch the seriousness of the challenge.

Maybe these books will be found some day and heal a few.

I have to always acknowledge those who make a difference to me, and who started it all. This is the short list. Ruth & Bud, my parents, for gifts I am still counting. And... for my little brother, Dennis, who reminded me when I was wrong.

Then there are my daughters, Candice and Courtney, who thought they knew their dad, but really didn't. There is my brilliant wife Christina, who thought she knew her husband..... And then there are my friends from the past whose life journeys I do not fully know, and who do not know me now. For in life it is who we become, not who we were.

So many people inhabit the fabric of our lives. So many played unseen roles in the writing of the eight books so far (only on Amazon):

1-800-I-AM-UNHAPPY Vol. 1

1-800-I-AM-UNHAPPY Vol. 2

1-800-FOR WOMEN-ONLY

1-800-LAUGHING-OUT-LOUD

1-800-OH-MY-GOODNESS

1-800-FOR-SEALS-ONLY

1-800-OH-MY-DONALD

And now, this the 8th is for all who know what veterans truly are.

Lastly, there are Sandra Simmons-Dawson and Brian Dawson who helped edit and format the books, website, and marketing. Their firm, Money Management Solutions, Inc. dba Customer Finder Marketing in Naples, FL http://customerfindermarketing.com/ is a gem.

God Bless America

IN THE WORDS OF OTHERS

Reviews for 1-800-I-Am-Unhappy (Volumes 1 and 2)

"This is a book by a man of many directions and passions. Straightforward yet thought provoking. Loyal to his convictions and country. And brave. Sharing. Warrior. Humanitarian."

Jeff Lytle, Editorial Page Editor, Naples Daily News

"As a friend, Chris has helped me understand the inherent conflicts embedded in the language of 'political correctness' and how it attempts, and frequently succeeds, in disguising and defeating the 'truth.' Chris is engaged in a rhetorical battle — we need his insight."

William Lord, a 32-year-veteran Executive Producer and Vice-President of ABC News, and Professor of Journalism at Boston University

"Chris writes like he lives. As a man of distinction, he is a voice for the poor, a champion of the truth and a friend of strong character and conviction. His word and his service are a blessing to all who encounter him."

Vann R. Ellison, President/CEO, St. Matthew's House, Inc.

"My nickname for Chris is "Dream-Catcher"- because that's who he is to me. He is my mentor in how to give on His behalf. Freely and generously, Chris offers both words, "God bless you!", and gifts. And all the while he is making a compelling and powerful statement. Chris Bent has discovered a beautiful way to live!"

Rev. Dr. Ruth Merriam, The Church on the Cape (U.M.C.), Cape Porpoise, Maine

"Chris Bent is a very unusual person – Navy SEAL, Yale graduate, successful business owner, and radical Christian who is comfortable talking with anyone at any level in society. He doesn't just talk about faith or caring about the poor, Chris actually lives his faith and he works with the poor. His smile is genuine and reflects his deep joy in life, America, hard work, people and (most definitely) God. I have enjoyed reading his writings; they are different, often hard hitting and sometimes maybe even a little wild. Each one gives a fresh perspective on contemporary lives, reflecting Chris' intel- ligence and faith. Chris enjoys moving mountains."

Rev. Dr. Ted Sauter, Senior Pastor, North Naples United Methodist Church

Reviews for 1-800-For-Women-Only

"It is amazing that a man would want to write about women. That is a change, but Chris has a sense of humor that can make you laugh. Women will enjoy this book and men may gain new insight."

Dorothy K. Ederer, O.P., Director of Campus Ministry,
St. John Student Center, East Lansing, Michigan

"Light, refreshing take on some not so light topics. Wrapped in silliness and wit are serious, social and moral truths that challenge us to be more than ordinary."

Peggy Ryba, Membership Director, North Naples Church, Naples, Florida

"Chris is like a modern day prophet, throwing modern day concepts and concerns out there for us to contemplate. The seeds he tosses can land on sand or soil depending on the reader. I suggest you pull up a nice spot in your garden and sit down and read…then allow some of his thoughts to germinate in your life! "

Mia Guinan, Owner, Gourmet Gang, Camp Trident, Virginia Beach VA

"Paradox is a person that combines contradictory features. Chris Bent is a paradox. Reading his most recent works I am not surprised by the depth, humor, passion and spirituality. In spite of mixed content the flow between chapters allows you to enjoy the paradox. Chris' muses have caused a few smiles; some ponderings and touched my heart. Let this Paradox of a Man walk up to you and continue the conversation.

Nancy Lascheid, RN, BSN, Co-Founder, Neighborhood Health Clinic, Naples, Florida

"1-800-For-Women-Only or the "Mystery of Women" is interesting because it is brutally accurate. In fact, it is frightening to read the explanations of characteristics of women. Many of these things I had not even been aware of, but they are "right on target". The book is written with great sensitivity and insight. I never got the feeling that women were criticized, but accepted as observed. It is an easy fun read and a great gift to give to a friend or even a son who is even thinking of getting married. As the mother of three sons, I know it is true; "Heart-felt is at the core of being. Being somebody."

Sue Lester, Volunteer, Children's Coalition of
Collier County, Pilot Club, Naples, Florida

"Chris Bent's extraordinary life has given him a perspective that so very few have. His insight comes not only from his incredible experiences but from his deeply rooted sense of responsibility, caring, and love for others. His thoughtful mind is not on idle, but instead always on overdrive, crystallizing in well thought out words those concepts that would have many times escaped us, were it not for the efforts of this author to engage, care deeply, and then, as Chris has done so remarkably here, write."

Jennifer L. Whitelaw, Attorney, Whitelaw Legal Group, Naples, FL

Reviews for 1-800-Laughing-Out-Loud

"Chis is a stew: meat, potatoes, veggies, gravy, biscuits and mustard. A warm, tender mix of good taste, generous servings, and something for all appetites! Chris mixes a Hunter S. Thompson "Gonzo Journalism" writing style with a Soupy Sales "Pie in the Face" sense of humor. Chris writes about: Life Values, Family, Self, Respect, Good & Evil. His perspective of life's Value Proposition engages our brain to think about ourselves and others. Chris' previous books are from the Heart and Soul. Take his counsel of his life's experience. There is good advice in each chapter! You will enjoy each word like every bite of a good stew."

Gerry Ross, Executive, Pratt & Whitney (Retired)

"Chris Bent is the type of guy you want to share a cold beer with at the end of a lousy day and have him philosophize on the real meaning of life. Since you might not have that opportunity anytime soon let me suggest you read 1-800-LAUGHING-OUT-LOUD. Perfect title for the book, because when reading it you will."

Nancy Lascheid, RN, BSN, Co-Founder, Neighborhood Health Clinic, Naples, Florida

Reviews for 1-800-Oh-My-Goodness

"With 1-800-Oh-My-Goodness, Chris Bent offers his thoughts on a variety of topics, in order to amuse, inspire, and challenge any reader. With his witty insight, and perspective forged from life experience, Chris seeks to help us all become better individuals."

Michael Hopkins, Attorney, Naples, FL

"In this book Chis is honest and open with the reader. He definitely gives you a lot to ponder. You can't wait to see what he is going to share next."

Dorothy K. Ederer O.P., Director of Campus Ministry, St. John Student Center

"Oh my goodness", Chris has again presented a faith filled and thought provoking book. His stream of thought, that often reads more like poetry than prose, will cause you to rethink moments of life in a context of love and promise."

Rev Jean Moorman Brindel, CFRE, AFP, Associate Director of Development,
Emeritus United Theological Seminary, Dayton Ohio

"Honest, incisive, poetic and profound: the writings of Chris Bent. Passion for people, the nation and the world spring from his pages; provocative questions leap from the shortest chapters ever. Silent voices speak in these pages and nothing is to be taken for granted, for life and love run deep between the lines of 1-800- Oh-My-Goodness. "

Wendy J. Deichmann, PhD, President, United Theological Seminary

Reviews for 1-800-For-SEALS-Only

"Pungent, cogent, wistful, idealistic, naive, wise, — all in no particular sequence, reflecting a view of life that it is all unpredictable, and it is mental, physical & moral preparation that will sustain us… there are life lessons and observations here for anyone and everyone…."

Lt (jg) James Hawes, BUDS 29E, SEAL, CIA, (He was the First SEAL In Africa)…(sadly was my UDTR Instructor too)

"Who knew SEALs could write? (LOL) But what Chris does with his gift is really less "writing" than it is expressing the "unwritten." We all have our thoughts; and Frogmen have certain very special and unique shared experiences. Chris puts the pen to the task of relating what we (the Frogs) have experienced and what we (all of his readers) now observe in sharing the experience of the world around us. It's challenging and funny (if you've been through a "real Hell Week"), and sometimes sad. But hey, isn't life? Hooyah!"

Timothy Phillips, SEAL, BUDS 166, ST-8, ST-4

"Chris - great stuff…as always. "Hooyah Mike"…"Every sin is a grenade"… "My wife is my swim buddy"…great thoughts as only a SEAL can put into words. I love it and will BUY a few copies for my Assistant Sergeant at Arms to read to guide their young lives… Hooyah Chris and see you soon!"

Phil King, Sergeant at Arms, NC Senate, BUDS 32

"Mr. Bent's words of wisdom on some of the evolutions of U. S. Navy SEAL training are demonstrated to apply to everyday life with such simplicity. God, Family, Country, is the essence of being an honorable and patriotic American. It is the ethos of the Navy SEAL credo. The band of brothers whose lives are bonded as one in being; all for one and one for all! Nothing in this world feels better to receive in life as the emblem, the SEAL Trident, of a true warrior and to receive into one's heart the holy trinity! Hooyah! The only easy day was yesterday!"

Erasmo Elijah Riojas (Doc Rio) HMC (SEAL) Ret.

"I am a SEAL Teammate of LT. Chris Bent. During our years of serving our country as Naval Special Warfare Operatives, Chris always manifested that "Can Do" attitude so necessary for success in what many would consider: "A tough way to make a living!"

Among other sub-specialties, Chris and I had the honor of being the Platoon Commanders who would "Recover Astronauts!" Within the pages of "1-800-FOR-SEALS-ONLY", you will get to see the mind-set of students going through BUDS Training (still the toughest Military Training in the World) with most Classes experiencing an over 80% Drop Out Rate! Chris masterfully combines our training to current issues existing today. A Giant HOOYAH for a must read publication! 1-800-FOR-SEALS-ONLY is awarded a big BRAVO ZULU from your old Teammates!"

Dr. Frank Cleary, OIC, Seventh Platoon, ST-2 (Ret.)

"One need only look into the night sky to recognize that there is brilliance in chaos. One need only read this book to realize the same. Intertwined in stories, random thoughts, and opinions one will find extraordinary pearls of wisdom in here..........and a lot of them. Chris is brilliant."

Navy SEAL Commander

"Dear Frogfather, Your writings remind me of the lessons and examples that were taught to me and my siblings by my parents, grandparents and the nuns that taught me in parochial school. I am so blessed to have them in my life. We are also blessed to have you because you have taken the time and effort to put down in writing your thoughts. They are insightful, and positive, to help us lead a better life. Thank you."

Maureen Murphy, Mother of LT. Michael Murphy,
Medal of Honor recipient, BUD/S Class 236, SDVT-1

"Five Stars for the FROGFATHER! This is a great book, and should be required reading...."

Commander (SEAL) Tom Hawkins, USN, Ret., author, NSW Historian

"Chris Bent has again taken his many and varied life experiences and applied them to life in general and "how to do it right". This book is clearly for everyone, not just SEAL's. Life was never meant to be easy and all of us can take away something from this book and the Frogman saying "The only easy day was yesterday". Even if it is the hard way....do the right thing.

From one Frogman to another I say to Chris, your eulogy (chapter 75) should be read when the time comes: Teammate, seen or unseen, you truly have made a difference!

Hooyah 1-800-For-SEALS-Only!"

Mike Macready, SEAL Team One, BUD/S 49 West Coast

"Chris Bent's latest 1-800 offering certainly gets my SEAL of approval... Using his own unique blend of insight, intellect and inspiration, Chris lifts parallels from the rich history and tradition behind the US Navy SEALs to provide challenging questions and equally thought provoking answers to this experience that we call life. In this social-networking, politically-corrected day and age where common sense, discipline and values seem to have fallen by the wayside, Chris Bent cuts through like a K-Bar to remind us all exactly what is of the utmost importance."

Darren A. Greenwell - NSW Historian, Researcher, Collector

Made in the USA
Charleston, SC
25 October 2015